T0139321

MEDICAL
INTELLIGENCE
UNIT

Immunophenotypic Analysis
Second Edition

Katalin Pálóczi, M.D., Ph.D., D.Sc.

Professor and Chair of Immunology
Faculty of Health Sciences
Semmelweis University
National Institute of Haematology and Immunology
Budapest, Hungary

LANDES BIOSCIENCE
GEORGETOWN, TEXAS
U.S.A.

EUREKAH.COM
AUSTIN, TEXAS
U.S.A.

IMMUNOPHENOTYPIC ANALYSIS
SECOND EDITION

Medical Intelligence Unit

Landes Bioscience
Eurekah.com

Please address all inquiries to the Publishers:
Landes Bioscience / Eurekah.com, 810 South Church Street, Georgetown, Texas, U.S.A. 78626
Phone: 512/ 863 7762; FAX: 512/ 863 0081
http://www.eurekah.com
http://www.landesbioscience.com
ISBN: 1-58706-101-5

While the authors, editors and publisher believe that drug selection and dosage and the specifications and usage of equipment and devices, as set forth in this book, are in accord with current recommendations and practice at the time of publication, they make no warranty, expressed or implied, with respect to material described in this book. In view of the ongoing research, equipment development, changes in governmental regulations and the rapid accumulation of information relating to the biomedical sciences, the reader is urged to carefully review and evaluate the information provided herein.

Library of Congress Cataloging-in-Publication Data

Pálóczi, Katalin.
   Immunophenotypic analysis / Katalin Pálóczi. -- 2nd ed.
      p. ; cm. -- (Medical intelligence unit)
   Rev. ed. of: Clinical applications of immunophenotypic
analysis. c1994.
   Includes bibliographical references and index.
   ISBN 1-58706-101-5
   1. Immunophenotyping. 2. Hematopoietic stem cell disorders
--Diagnosis.   I. Pálóczi, Katalin. Clinical applications of im-
munophenotypic analysis. II. Title. III. Series: Medical intelli-
gence unit (Unnumbered : 2003)
   [DNLM: 1. Immunophenotyping. 2. Bone Marrow--immuno-
logy. 3. Hematopoietic Stem Cells--immunology. 4. Leukemia
--immunology. 5. Lymphoma, Non-Hodgkin--immunology.
QY 250 P1818i 2004]
   RC644.5.P25 2004
   616.07'582--dc21

                              2001004704

To our family members whose patience
and unwavering support have allowed us to share
important new information with those dedicated
to the conquest of malignant disease

CONTENTS

EDITOR

Katalin Pálóczi, M.D., Ph.D., D.Sc.
Professor and Chair of Immunology
Faculty of Health Sciences
Semmelweis University
National Institute of Haematology and Immunology
Budapest, Hungary
E-mail: k.paloczi@ohvi.hu
Chapters 4, 5

CONTRIBUTORS

Angela Dolganiuc, M.D., Ph.D.
Department of Medicine
University of Massachusetts Medical School
Worcester, Massachusetts, U.S.A.
Chapter 1

András Matolcsy, M.D., Ph.D., D.Sc.
Associate Professor
Department of Pathology
Faculty of Medicine
Pecs University
Pecs, Hungary
E-mail: amatolc@pathology.pote.hu
Chapter 3

Nóra Regéczy, M.D., Ph.D.
Membrane Biology
and Immunopathology Research Group
Hungarian Academy of Sciences
National Institute of Haematology
and Immunology
Budapest, Hungary
E-mail: nregeczy@hotmail.com.hu
Chapter 2

Gyongyi Szabo, M.D., Ph.D.
Professor
Department of Medicine
University of Massachusettes Medical School
Worcester, Massachusetts, U.S.A.
E-mail: gyongyi.szabo@umassmed.edu
Chapter 1

PREFACE

The first edition of *Clinical Application of Immunophenotypic Analysis* appeared in 1994. This work was intended to be a comprehensive and up-to-date book developed to serve a diverse group of individuals ranging from practicing general immunologists, hematologists, students of the field, basic scientists involved in research of cell surface antigens, as well as practicing internists and pediatricians. The science and practice of flow cytometry and immunophenotyping have continued to evolve. With more and more clinical information and evaluation demonstrating the value of immunophenotypic analysis, this has become an established clinical procedure.

The rapid advancement of this field has necessitated the creation of a second edition, in order that the book remains current and useful to our readers. Besides, there is a need to put a great deal of effort into promoting the value of clinical flow cytometric tests which can be more efficiently performed and provide more definitive results than many of the classical procedures. This is especially important in areas of clinical immunology and hematology.

We retained the features of the first edition that were regarded by our audience as most valuable, but altered the second edition as required to further enhance the quality of the book. We have incorporated much new information regarding normal immune cell development as well as diagnosis of acute leukemias, malignant lymphomas and characterization of stem cells for transplantation. To accommodate these new insights without significantly increasing the size of the book, we have reduced discussions of historical debates that are now resolved by experiments.

All chapters have been rewritten and extensively updated to reflect the explosion of knowledge. Beside the editor four excellent contributors have been involved in creating the second edition and all the authors are working leaders in their subjects. Most of the authors confront their clinical problems on a daily basis and have that uniquely urgent spur to keep them at the cutting edge. As a result of the collective labor of a creative people, we now have a book, which hopefully fills a real need in one of the most exciting branches of modern medicine.

The contributors to this edition have done an outstanding job, and the editor is indebted to them for their diligence, perseverance, and scholarly presentations.

The Editor

ABBREVIATIONS

ADA	adenosine deaminase
ADCC	antibody dependent cytotoxic-cytolytic activity
Ag	antigen
AGM	aorta-gonad mesonephros
AIDS	acquired immunodeficiency syndrome
AL	acute leukemias
ALCL	anaplastic large cell lymphoma
ALL	acute lymphoid leukemia
AML	acute myeloid leukemia
ATL/L	adult T cell leukemia/lymphoma
AUL	acute undifferentiated leukemia
BCR	B cell receptor
BL	Burkitt's lymphoma
B-LBL	B-lymphoblastic leukemia/lymphoma
BM	bone marrow
BMT	bone marrow transplantation
CALLA	common acute lymphoid leukemia antigen
CD	cluster of differentiation
CGD	chronic granulomatous disease
CLL/SLL	chronic lymphocytic leukemia/small lymphocytic lymphoma
CTL	cytotoxic T lymphocytes
DCs	dendritic cells
DLBL	diffuse large B cell lymphoma
EBV	Epstein Barr virus
ECM	extracellular matrix
FAB	French-American-British
FL	follicular lymphoma
GM-CSF	granulocyte-macrophage colony stimulating factor
GVHD	graft versus host disease
HCL	hairy cell leukemia
HHV8	human herpes virus 8
HIV	human immunodeficiency virus
HLA	human leukocyte antigen
HPC	hematopoietic progenitor cell
HTLV-1	human T cell lymphotropic virus-1
IFN	interferon
Ig	immunoglobulin
IL	interleukin
KIR	killer inhibitory receptors
KSHV	Kaposi's sarcoma associated herpes virus
LPL/IC	lymphoplasmocytoid lymphoma/immunocytoma

MALT	mucosa associated lymphoid tissue
MCL	mantle cell lymphoma
MDR	multidrug resistance
MDS	myelodisplastic syndrome
MF	mycosis fungoides
MoAb	monoclonal antibody
MPO	myeloperoxidase
MRD	minimal residual disease
MZL	marginal zone lymphoma
NHL	non-Hodgkin's lymphoma
NK	natural killer
PBPC	peripheral blood progenitor cell
PCR	polymerase chain reaction
PLL	prolymphocytic leukemia
PNP	purin nucleotide phosphorilase
REAL	Revised European American Lymphoma classification
SCID	severe combined immunodeficiency
sm	surface membrane
SS	Sezary syndroma
SSC	side scatter
TCR	T cell receptor
Tdt	terminal deoxyribonucleotydil transferase
T-LBL	T-lymphoblastic leukemia/lymphoma
TNF	tumor necrosis factor
T-PLL	T-prolymphocytic leukemia
WAS	Wiskott Aldrich syndrome
XLA	X-linked agammaglobulinemia

Acknowledgements

Landes Bioscience has provided excellent support
for this project and we thank the team of publishers
for their continued confidence in our efforts.
In particular, Ronald G. Landes and Cynthia Conomos
are acknowledged for their extraordinary contributions.

CHAPTER 1

Normal Hematopoietic Cell Differentiation

Gyongyi Szabo and Angela Dolganiuc

Immunologic Marker Analysis of Cells in the Various Stages of Hematopoietic Differentiation

Differentiation Pathways of Normal Hematopoietic Cells

Hematopoiesis is a complex process in which hematopoietic stem cells self-replicate and differentiate to generate the different mature cell types in the blood and lymphoid organs. Under steady-state conditions, stem cells and hematopoietic progenitor cells reside in the bone marrow (BM) medullary cavity in contact with the bone marrow microenvironment, which includes a diversity of mesenchymal cells that secrete hematopoietic cytokines and extracellular matrix components. Most of the factors required for the orderly development of stem cells are present in the BM microenvironment. More than 40 different growth factors, cytokines, and chemokines interact with stem and progenitor cells through specific receptors and regulate proliferation, differentiation, and cell fate. Hematopoietic growth factors are produced by both mesenchymal and hematopoietic cells and can be present in soluble, cell-bound forms, or bound to the extracellular matrix (ECM).[1,2] Stem and progenitor cells express adhesion molecules that provide specific cell-cell and cell-ECM interactions. In addition, cytokines and growth factors affect adhesive interactions between stem and progenitor cells and their adhesive ligands in the BM providing a further level of regulation.

During development, cells express various surface markers that serve as a tool in their identification (phenotypes).[3-5] These molecules, recognized by various monoclonal antibodies, were defined as Clusters of Differentiation (CD) for both standardization procedures and immunological analysis.[6-8] While some CDs appear temporarily during different stages of cell differentiation, others are present during the entire life-span of a particular cell; some CD are specific only for one differentiation stage and/or cell lineage, others are common for multiple differentiation stages of a particular lineage or could recognize cell of multiple lineage descendents. The number of characterized CDs has expanded during the last few years and increasing number of specific monoclonal antibodies are offered by different laboratories.

Pluripotent stem cell can differentiate into lymphoid and myeloid progenitors (Table 1).[9,10] Within the lymphoid differentiation B, T, NK and NKT cells lineage can be distinguished. The myeloid progenitors give rise to monocytic, granulocytic, erythroid and megakaryocytic differentiation.[10] Dendritic cells develop from either lymphoid or myeloid precursors.[11]

Table 1.   Surface Ag expressed during hematopoiesis

Lymphoid Stem Cell		Stem Cell CD34+ Myeloid Stem Cell			
Pro-B cell HLA-DR CD19 (CD22) Cytoplasmic TdT	**Early thymocyte** CD2 (CD3) CD7 TdT	**Myeloblast** HLA-DR CD13 CD33	**Monoblast** HLA-DR CD13 CD14 CD33	**Megakaryoblast** HLA-DR CD41 CD42 CD61	**Erythroblast** HLA-DR CD71 Glycophorin A and C
Pre-B cell HLA-DR CD10 CD19 CD20 CD22	**Intermediate thymocyte** CD1 CD2 CD3 CD4 CD5 CD7 CD8 TdT	**Promyelocyte** CD13 CD15 CD33	**Promonocyte** HLA-DR CD13 CD14 CD33		**Normoblast** CD71 Glycophorin A and C
Immature B cell HLA-DR CD10 CD19 CD20 CD21 CD22 Surface Ig	**Mature thymocytes** CD2 CD3 CD4/CD8 CD5 CD6 CD7	**Myelocyte** CD13 CD15 CD33		**Megakaryocyte** CD41 CD42 CD61	
Mature B cell HLA-DR CD19 CD20 CD21 CD22 CD23 CD37 Surface Ig	**Mature T cell** CD2 CD3 CD5 CD6 CD7 CD4/CD8 CD25 (activated T cells) CD43	**Neutrophil** CD11b CD13 CD15 CD33 CD43	**Monocyte** HLA-DR CD11b CD11c CD13 CD14 CD33	**Platelet** CD41 CD42 CD61	**Erythrocyte** Glycophorin A and C

Lymphoid Descendents

Populations of B Cells

B cells produce antibodies, the cornerstones of "humoral immunity", and participate in immune memory. B cell differentiation occurs partly in the bone marrow (antigen independent differentiation) and partly in the peripheral lymphoid tissues (antigen-dependent proliferation) (Table 2).[12] Two different subpopulations of B lymphocytes, B1 and B2 cells have been characterized.[13]

Table 2. Human B cell development and markers

- Pre-B-I cells: $CD19^+$ $CD10^+$ $CD34^+$ TdT^+ $RagI^+$ $VpreB^+$ $H\mu$-cycling cells have DJ_H-rearranged IgH-chain loci, while their L-chain loci remain in germline configuration. Defective pre-BCR expression arrests B-cell development at the transition from pre-B-I to large pre-B-II cells.
- Large, pre-BCR$^+$ pre-B-II cells: $CD19^+$ $CD10^+$ $CD34^-$ TdT^- $Rag1^-$ $VpreB^+$ $H\mu^+$ pre-BCR$^+$ cycling cells are productively V_HDJ_H rearranged in the IgH chain locus, while their L-chain gene loci remain in germline configuration.
- Large pre-BCR$^-$ pre-B-II cells: $CD19^+$ $CD10^+$ $CD34^-$ TdT^- $Rag1^-$ $VpreB^-$ $H\mu^+$ pre-BCR$^-$ cycling cells have their IgH-chain locus productively V_HDJ_H rearranged, while their L-chain loci are largely in germline configuration.
- Small resting pre-B-II cells: $CD19^+$ $CD10^+$ $CD34^-$ TdT^- $Rag1^+$ $VpreB^-$ $H\mu^+$ $sIgM^-$ small resting cells have a majority of their L-chain gene loci in V_JJ_L rearranged configurations but do not yet express L-chains on their surface.
- Immature B cells: $CD19^+$ $CD10^+$ $sIgM^+$ $sIgD^-$
- Mature B cells: $CD19^+$ $CD10^-$ $sIgM^+$ $sIgD^+$

Abbreviations: BCR= B-cell receptor; Rag1= recombinase-activating gene 1; TdT= terminal deoxynucleotidyl transferase

The B cells that develop earliest during ontogeny are referred to as B1 cells. Most B1 cells express CD5, an adhesion and signaling cell-surface molecule[13-15] and are the source of the natural antibodies, which can be produced without exposure to any environmental antigen or immunization. B1 cells are predominantly localized in the peritoneal cavity where they can be distinguished as IgM^{hi} IgD^{lo}, $CD9^+$, $IL-5R^+$, $CD23^-$ $CD43^+$ and $CD5^+$ (B-1a) or $CD5^-$ (B-1b) cells.[14]

The splenic B cell population lacks the CD5 molecule, and because these cells develop slightly later in ontogeny, they are referred to as B2 cells. Before they encounter antigen, mature B2 cells co-express IgM and IgD antibodies on their cell surface, but by the time they become memory cells, they have usually switched to the use of IgG, IgA, or IgE as their antigen receptors. Marginal zone B cells express CD9 but not CD5, IL-5R or CD43 and play an important role in T cell independent antibody response via activation of Pyk-1 tyrosine kinase.[15] Association of CD21 with CD19 and B cell receptor (BCR) lowers activation threshold of naïve B cells.[16] The final stages of differentiation of B2 cells into antibody-secreting plasma cells occur within the secondary lymphoid tissues outside the germinal centers. Although generally short-lived, with a half-life of only a few days, some plasma cells survive for weeks, especially within the bone marrow. The CD markers associated with B cells are summarized in Table 3 (for additional details see ref. 8).

Populations of T Cells

T lymphocytes are the major effectors of "cellular immunity", produce cytokines, provide signals for B cells differentiation, act as memory and regulatory compartment of immunity. The antibody-dependent cytotoxic-cytolitic (ADCC) activity and cytotoxic T lymphocytes (CTL) is key to tumor immunity. Adult peripheral T cells developed from bone marrow derived thymic lymphoid progenitor.[9] T cell development in the thymus can be subdivided into two main phases that govern lineage fate decision: an early T cell developmental stage before CD4 and CD8 coreceptor expression, in which a common T-lineage-restricted precursor gives rise to $\alpha\beta$ or $\gamma\delta$ T lineage cells; and the T cell selection phase in which $\alpha\beta$ lineage $CD4^+CD8^+$ double positive (DP) cell commit to become either $CD4^+$ helper of $CD8^+$ cytotoxic T cells.[17-19]

Table 3. B cells

Receptor	Family	Molecular Weight (kDa)	Distribution	Function
BCR: The BCR is a complex of surface membrane Ig (sIg) and the associated proteins CD79a and CD79b.				
sIg	Ig-SF prototype	26-73	On all mature B cells and most neoplastic B cells	Specific binding to antigen
The surrogate BCR complex	Ig-SF	20	Pro- and pre-B cells	Not known: probably important for early B-cell proliferation and allelic exclusion. The natural ligand has not been identified.
MHC class II	Ig-SF related	35	B cells, monocytes, macrophages, myeloid and erythroid precursors, DCs, some epithelial cells	Presentation of processed antigen peptides to CD4$^+$ T cells. Role in positive and negative selection of CD4$^+$ T cells in thymopoiesis.
CD5	SRCR-SF	67	Subset of mature B cells, mature T cells, thymocytes	Unknown. CD5$^+$ B cells (B-1) cells implicated in autoimmune disease.
CD9	TM4-SF	22-27	Early B cells, activated B cells, activated T cells, platelets, eosinophils, basophils and endothelial cells	Platelet activation and aggregation, cell-cell adhesion (pre-B cells). Possible role in signaling mediated by interaction with GTP-binding proteins.
CD10 CALLA	Zinc metallo-protease	100	Widely expressed, including early T- and B-cell precursors, marker for CALL, neutrophils, fibroblasts, epithelial cells	Zinc metalloprotease. May act to limit activity of fMLP peptide hormones. Possible role in B-cell development.
CD19	Ig-SF	95	All human B cells and B-cell precursors, FDCs	Regulation of B-cell activation and proliferation. Part of signal transduction complex that includes CD81, CD21 and Leu-13. Reduces BCR signaling threshold.
CD20	Unassigned	33-37	All mature B cells, human pre-B cells, not plasma cells	Can generate intracellular Ca^{2+} signal, may be important for regulation of B-cell activation and proliferation.

Table continued on next page

Table 3. Continued

Receptor	Family	Molecular Weight (kDa)	Distribution	Function
CD21	Member of RCA gene cluster	145	Mature sIg$^+$ B cells (lost on activation), FDCs, subset of normal thymocytes	Receptor for C3d, EBV. Role in B-cell activation and part of signaling complex that includes CD19, CD81 and Leu-13. Binds CD23 and is implicated in regulation of IgE production.
CD22	Ig-SF	130-140	Surface expression on mature B cells, cytoplasmic expression in late pro- and early pre-B cells. Lost prior to plasma-cell stage.	Mediates adhesion to erythrocytes, T cells, B cells, monocytes and neutrophils. CD22-deficient mice have increased sensitivity to BCR activation, thus probably limits BCR signal.
CD23	Novel superfamily of type II proteins with C-type lectin motifs	45	FcɛRIIa form expressed on mature B cells and monocytes, FcɛIIb on monocytes, IL-4 activated macrophages, eosinophils, platelets, DCs, FDCs	Low affinity IgE receptor: role in IgE regulation. Binds CD21 and has role in B-cell activation, may also prevent apoptosis of germinal center B cells.
CD24	CD52/CD24/HAS	35-45	Expressed throughout B-cell development. Decreased on activation and lost on plasma cells. Present on mature granulocytes and on epithelial cells.	Play a role in regulation of B-cell growth and antibody production
CD25 IL-2Rα	New family with IL-15R α-chain: has homology with CCP	55	Activated B cells, T cells, monocytes, NK cells and macrophages	Induces activation and proliferation of T cells, B cells, thymocytes, NK cells and macrophages
CD32 FcγRII	C2 Ig-SF	40	FcγRIIB form present on B cells, FcγRIIA and C forms expressed on neutrophils; all forms are found on monocytes. FcγRII proteins are also on endothelial cells.	Low-affinity receptor for aggregated IgG. Binding to FcγRII on B cells can deliver a negative signal to B cells.
CD37	TM4-SF	40-52	Mature normal and neoplastic B cells; neutrophils and monocytes; low expression on T cells	Role in signal transduction and intracellular trafficking

Table continued on next page

Table 3. Continued

Receptor	Family	Molecular Weight (kDa)	Distribution	Function
CD40	TNFR-SF	48-50	Normal and neoplastic B cells, FDCs, IDCs, endothelial cells, thymic epithelium, macrophages	Binds CD40L (CD154) on activated T cells. Key signal for B-cell activation, proliferation and differentiation, formation of GCs, isotype switching, rescue from apoptosis and differentiation into memory or plasma cells.
CD45	PTP	220	All haematopoietic cells	PTP activity. Has critical regulatory role in TCR and BCR activation.
CD72	C-type lectin	43	All B cells except plasma cells; spleen red pulp macrophages, Kupffer cells	Induce B-cell activation and proliferation
CD79a CD79b	Ig-SF	33-39	Normal and neoplastic B cells, expressed early during B-cell ontogeny	CD79 forms noncovalent association with sIg and is important for BCR signal transduction. CD79 may also mediate transport of IgM to cell surface.
CD80	Ig-SF	60	Activated B cells, activated T cells, IFN-γ-stimulated monocytes, DCs	Early activation marker. Regulates IL-2 gene expression and T-cell activation.
CD81	TM4-SF	26	All B cells and T cells, leukaemia and lymphoma lines, neuroblastoma and melanoma lines, GC FDCs, NK cells, thymocytes, eosinophils	Part of signal transduction complex that includes CD19, Leu-13 and CD 21. Cross-linking CD81 induces effects consistent with a role in signal transduction, e.g., homotypic adhesion.
CD83	Ig-SF	43	Weakly expressed on activated B cells, GC B cells and lymphoblastoid B-cell lines. Langerhans cells, blood DCs and interdigitating reticulum cells.	Has role in antigen presentation/ cell-cell interactions following activation
CD86	Ig-SF	80	Activated B cells, blood monocytes, mature circulating B cells, DC	Expressed rapidly following B-cell activation. Regulate IL-2 expression and prevent T-cell anergy.

Table continued on next page

Table 3. Continued

Receptor	Family	Molecular Weight (kDa)	Distribution	Function
CD95	TNFR-SF	45	Expressed at high levels on activated B cells and T cells	Fas ligation induces apoptosis. Fadd/MORT1 associates with intracellular death domain.
CD122	CKR-SF	70-75	B cells, T cells, NK cells, monocytes, macrophages	Signaling subunit of IL-2R and IL-15R
CD124 IL-4R	CKR-SF	130-150	Mature B and T cells, haematopoietic precursors, pre-B, B- and T-cell lymphomas	Activates B cells causing increased expression of surface CD23 and IgM. Stimulates proliferative activity in pre-activated B and T cells. Switch factor involved in IgE regulation.
CDw125 IL-5Rα	CKR-SF	60	Eosinophils, basophils	Promotes growth and differentiation of eosinophil precursors and activates mature eosinophils
CD126 IL-6R	CKR-SF	80	Activated B cells, plasma cells, T cells, monocytes, epithelial cells, fibroblasts hepatocytes, neural cells	Promotes growth of myelomas, B-cell hybridomas, activated and EBV-transformed B cells and T-cell lines. Induces differentiation and proliferation of haematopoietic precursors and mediates acute phase response of hepatic cells.
CD127 IL-7Rα	Low homology with CKR-SF	68	Thymocytes, T-cell and B-cell progenitors, mature T cells, monocytes, some lymphoid and myeloid cell lines	Stimulates proliferation of pro- and pre-B cells, thymocytes and mature T cells, and induces monocyte activation
CD132 IL-2Rγ common gamma chain (γc)	CKR-SF	65-70	B cells, T cells, NK cells, monocytes, macrophages, neutrophils	Signaling component receptors for IL-2R, IL-4R, IL-7R, IL-9R, IL-15R
IL-10R	Class II CKR-SF	110	B cells, thymocytes, mast cell and macrophage cell lines	Induces B-cell proliferation and differentiation, switch factor for IgA secretion with CD40L and TGF-β

Table continued on next page

Table 3.   Continued

Receptor	Family	Molecular Weight (kDa)	Distribution	Function
IL-13Rα	CKR-SF	49	Human B cells, endothelial cells, other nonhaematopoietic cells	Activates STAT6, induction of inflammatory cytokines
IL-15Rα	New family with CD25	65-75	mRNA found in T-cell lines, macrophage lines, bone marrow stromal cell lines. Especially abundant in liver.	Binds IL-15 with high affinity. Complexes with IL-2R β-chain and γc-chain to form IL-15R complex. Soluble form identified.

During thymic development, these precursors expand by a million fold, differentiate into two T cell and at least two non-T cell lineages, and, within the T cell lineage, acquire the capacity to express the Ag-specific TCR complex. The two T cell lineages are designated alpha/beta and gamma/delta, depending upon expression of the respective TCRs.[17-20] Expression of the alpha-, beta-, gamma-, and delta-chains of the TCR requires somatic recombination of the V, D, and J genes encoding the V domain of the corresponding TCR proteins.

Stage 1 precursors can give rise to multiple lineages, including T, NK, and dendritic cells (DCs), but not B cells or myeloid cells. The successive steps of immature, CD4⁻/CD8⁻ thymocyte (double negative) differentiation have been characterized by surface expression of CD44 in the absence (stage 1) or presence (stage 2) of CD25 followed by an initial down-regulation of CD44 (stage 3) and subsequently of CD25 (stage 4). Stages 1 and 2 have been shown to have the TCR-beta and TCR-gamma loci in germline configuration. Both partial (D to J) and complete (V to DJ) rearrangements of the TCR-beta locus as well as V to J rearrangements of the TCR-gamma locus occur primarily in stage 3. No further rearrangements of these loci (beta, gamma and delta) are known to occur after the transition to stage 4, which marks the onset of TCR locus recombination and commitment to the alpha/beta lineage.[17-21]

T helper and T cytotoxic/suppressor lymphocytes can be divided into two distinct subsets of effector cells based on their functional capabilities and the profile of cytokines they produce.[22,23] The Th1 subset of CD4⁺ T cells, as well as Tc1 subset of CD8⁺ cells secretes cytokines usually associated with inflammation, such as IFN-gamma and TNF and induces cell-mediated immune responses. The Th2 and Tc2 subsets produces cytokines such as IL-4, IL-5 and IL-10 that help B cells to proliferate and differentiate and are associated with humoral-type immune responses. The selective differentiation of either subset is established during priming and can be significantly influenced by a variety of factors. The separate Th0 subset was characterized by capacity to produce IL-2, IL-4, IL-5, IL-10, IFN-gamma and TNF and it is considered as a precursor for Th1 or Th2 development.[22,24,25] T cells require two signals for activation. The first signal is provided through the TCR and a second one via costimulatory molecules and/or signals.[26] The identification of a newly extended family of molecules related to the T-cell costimulatory proteins, CD28 and B7, suggests that a distinct form of costimulatory signals could be important for effector T cell responses.[27] The CD classification of markers and receptors of T lymphocytes are summarized in Table 4 (additional details are in ref. 8).

NK Cells

NK cells, like other hematopoietic cells, are derived from pluripotent stem cells. Evidence for a restricted NK/T cell progenitor came initially from analysis of early fetal thymocytes,

Table 4. T cells

Receptor	Family	Molecular Weight (kDa)	Distribution	Function
TCR: The TCR is a complex of surface TCR αβ or γδ and associated CD3 proteins (see below).				
TCRαβ and TCRγδ	Ig-SF	40-45	αβ on T cells; γδ in fetal and adult thymus, and the thymic independent population of the intestinal epithelium and other epithelia (skin, mucosa).	T-cell antigen-specific receptor for MHC-peptide complexes on APCs.
CD1	MHC-like	50-67	Cortical thymocytes	Peptide presentation, Lipid presentation
CD2 LFA-2, T11	Ig-SF	55-60	Thymocytes, mature T cells, murine B cells	Adhesion (CTLs to target cells, T cells to endothelium, and monocytes and thymocytes to TECs). Signal transduction; T-cell activation. Interaction with tubulin and regulation of anergy.
CD3	Ig-SF	Three chains: γ [25-28 (human), 21 (mouse)], δ (21, 28), ε (20, 25) associated to ζ (16) and η (22, 21) by disulphide bridges and to the TCR αβ chains.	All T cells	Signal transduction (ζ), T cell activation and function
CD4	Ig-SF	60	Thymocytes, Th cells, monocytes, bone marrow cells	Helper activity, coreceptor with the TCR and signaling in association with $p56^{lck}$; adhesion (stabilizes TCR-MHC class II interaction)
CD5	SRCR-SF	67	Thymocytes, mature T cells	T-cell costimulation. Thymocyte activation. Anti-CD25 stimulates T-cell proliferation, IL-2 secretion and an increase in Ca^{2+} concentration.
CD7	Ig-SF	40	Thymocytes, mature T cells	Signal transduction

Table continued on next page

Table 4. **Continued**

Receptor	Family	Molecular Weight (kDa)	Distribution	Function
CD8	Ig-SF	34	Thymocytes, CTLs, IELs (CD8αα), some DCs	Maturation and positive selection of MHC class I restricted CTLs
CD16	Ig-SF	50-65	Subsets of mature T cells, monocytes, B cells, NK progenitors and NK cells	ADCC. Thymocyte development.
CD25	Hemato-poietin receptor SF	55	Immature TN, activated T cells, pre-B cells	T-cell growth, enhances NK-cell activity
CD26	Ecto-peptidase IV	220	Thymocytes, T cells, NK cells	T-cell triggering, binds and transports ADA to the cell surface, binds to fibronectin, collagen
CD28	Ig-SF	44	T cells, resting T cells, plasma cells	Costimulatory molecule initiating signal transduction
CD30	TNFR-SF	105	Activated T and B cells	Transduction of a death signal
CD38 T10	Unassigned	46	Thymocytes, activated T cells, NK and DC precursors	Signal transduction. Cell adhesion.
CD43 gp115	Unassigned	115	T cells, induced on activated CD4+ T cells, pre-B cells	Leucosialin, Sialophorin. Role in T-cell proliferation, costimulation and adhesion.
CD45 LCA	Unassigned	180-220	Fetal liver T cell progenitors, mature, activated, naïve T cells (CD45RO), and memory (CD45RA) T cells	A protein tyrosine phosphatase that regulated Src-family kinases. Role in thymocyte development, selection and TCR-mediated signal transduction.
CD56 NCAM	Ig-SF	200-220	T cells, IL-2 dependent T-cell clones, NK cells	Induces killer activity and non-MHC restricted cytotoxicity
CD69 VEA	C-type lectin SF (group V)	85	Marker of positively selected thymocytes; T-cell precursors and TN thymocytes	Signal transduction in T cells. Apoptosis in eosinophils.

Table continued on next page

Table 4. *Continued*

Receptor	Family	Molecular Weight (kDa)	Distribution	Function
CD90 Thy1	Ig-SF	25-35 kDa	T-cell lineage, haematopoietic stem cells, pre-B cells	Role in lymphoycte recirculation, adherence, T-cell activation, cellular recognition
CD95 Fas, APO-1	TNFR-SF	36-45	Thymocytes, mature T cells, activated lymphocytes	Induces apoptosis signal when triggered by FasL or anti-Fas Ab. Role in clonal deletion of peripheral T cells and AICD of mature T cells.
CD100	Semaphorin	150	Normal PBLs, resting and activated T cells, most hematopoietic cells	Proliferation and activation of PBMCs. Role in T-cell homotypic adhesion and B-cell differentiation.
CD117	Ig-SF and receptor kinase	145-150	Hematopoietic progenitors	Signal transduction. Cell differentiation. Regulation of adhesion.
IL-2R α chain (CD25) β-chain (CD122), γ-chain (γc) (CD132)	CKR-SF	β (70) γc (64)	Activated T cells (α,β,γc). γc is constitutively expressed on lymphocytes; βγ on NK cells and CD8$^+$ T cells. CD4$^+$CD25$^+$ autoimmune-preventative T cells.	Responsible for all biological effects induced by IL-2. Increases cytolytic activity of NK cells and Ig biosynthesis by B cells. Signaling via γc can prevent induction of anergy in T cells.
CD124 IL-4R gp140	CKR-SF		T cells, B cells, haematopoietic precursors	Stimulates T-cell growth and B-cell activation to promote T-B interactions. Induces differentiation of Th2 cells: stimulates CTL development; induces CD8$^+$ cells produce IL-4.
CD127 IL-7R	CKR-SF	75	Immature thymocytes, mature T cells, human bone marrow lymphoid progenitors, γδ IEL	Induction and promotion of immature T-cell growth, pre-B/B-cell proliferation. Induces CD25 on T cells.
CD129 IL-9R	CKR-SF	64	T cells, B cells, macrophages, megakaryoblasts	Growth promoting activity for T-cell tumours. Inhibits apoptosis induced in thymic lymphomas.

Table continued on next page

Table 4. Continued

Receptor	Family	Molecular Weight (kDa)	Distribution	Function
IL-12R	CKR-SF	β1(100), β2(130); high-affinity receptor (β1β2)	CD4$^+$, CD8$^+$ activated T cells, human Th1 clones, NK cells, B-cell lines	Signal transduction (β2). Supports proliferation of T cells and NK cells.
IL-17R	Unassigned	98 (predicted)	Widespread, in particular expressed in TN thymocytes, T-cell lines (EL4) and T-cell clones	Enhances PHA-induced T-cell proliferation: sIL-17R inhibits T-cell proliferation and IL-2 production
CD152 CTLA-4	Ig-SF	70% sequence similarity with CD28	Activated T cells	Binds CD80 and CD86. Costimulatory molecule.
CD154 CD40L	TNF-SF	33	Activated CD4$^+$ T cells	Costimulates proliferation and lymphokine secretion from T cells. Activates and induces B-cell proliferation and differentiation.

many of which—like NK cells—express the Fcγ receptor III (FcγRIII). This FcγRIII$^+$ fetal thymocyte population gave rise to TCR$^+$ T cells after intrathymic transfer or to NK cells after intravenous transfer but are incapable of giving rise to myeloid cells or B cells. NK cell development normally occurs extrathymically, posing the question of whether restricted NK cell progenitors exist in the periphery.

In humans the well-established lineage characteristic for NK cell development is not known yet. In mice the FcRIII$^+$ fetal thymic population is heterogeneous. A fraction of the cells expresses the NK cell markers NK1.1 and DX5 but fail to express CD117 (c-kit)—NK1.1$^+$ DX5$^+$ CD117$^-$ cells—and another fraction exhibits the NK1.1$^+$ DX5$^-$ CD117$^+$ phenotype.[28] The CD117$^-$ subset exhibits ex vivo cytolytic activity against some tumor target cells, suggesting that it contains functionally mature NK cells. The CD117$^+$ subset is capable of reconstituting T cell development in fetal thymic organ culture and NK cell development when cultured in special conditions, but failed to give rise to B cells or myeloid cells. The CD117$^-$ population cannot give rise to T cells. Possibly related is the finding that human precursor thymocytes express NKR-P1A, a human isoform of the murine NK1.1 antigen. It was concluded that the CD117$^+$ population represents restricted progenitor cells for the T and NK lineages while the CD117$^-$ population represents mature NK cells.[28,29]

Human natural killer cells express the IL-15R chain, the IL-2/-15R chain and the common-cytokine-receptor chain (c), but do not express the whole IL-2R. IL-15 is necessary for both in vivo and in vitro NK cells differentiation from lymphoid-cell-restricted bone marrow progenitor-cell population.[29,30]

Human natural killer cell subsets can be distinguished by CD56 surface density expression.[31] In killing of cellular targets, NK cells employ receptors that activate them and receptors

Table 5. Human NKT cell markers

Characteristic	Human	Comment
Major subsets	CD4⁺, DN	Proportions vary
T cell receptor		
α-chain	Vα24JαQ	Homologous
β-chain	Vβll	Homologous
Expression level	Intermediate	
Accessory molecules		
NK associated	NKR-Pl	Homologous (CD161)
Restriction element	CDld	Homologous
Cognate antigen	Glycolipid	α-GalCer stimulates
Cytokine production		
IL-4	Rapid, high levels	Following TCR ligation
IFN-γ	+	Following TCR ligation
Frequency		
PBL	~0.1-0.5%	More variable in humans

Abbreviations: α-GalCer= α-galactosylceramide; DN= double negative; IFN-γ= interferon γ; IL-4= interleukin 4; NKR= NK-cell receptor; NKT= NK1.1⁺ T cells; PBL= peripheral blood leukocytes; TCR= T-cell receptor.

specific for MHC class I that inhibit their activation. There has been a rapid progression in recent years in the understanding of the inhibitory receptors that fall into two distinct structural types (CD94/NKG2 and killer inhibitory receptors [KIR]) that appear to utilize the same inhibitory signaling cascade (immunoreceptor tyrosine-based inhibitory motifs: ITIMs).[32]

NKT Cells

NKT cells are a phenotypically and functionally diverse cell population found in spleen, liver, bone marrow and thymus that are derived from lymphoid descendents of stem cells. They co-express some NK receptors such as NK1.1 (CD122, CD161 or NKR-P1) and T-cell receptor (TCR) (Table 5). The key features of NKT cells include heavily biased TCR gene usage with rearrangements of the AV24-AJ18 gene, CD1d restriction and high levels of cytokine production following TCR ligation, particularly for interleukin 4 (IL-4) and interferon gamma (IFN). NKT cells can be double-negative (DN) (CD4⁻/CD8⁻) or single-positive: CD4⁺/CD8⁻ or CD8⁺/CD4⁻.[33] Most of CD4⁺ NKT cells were found in thymus, while CD8⁺ NKT cells are preferentially extrathymic. In contrast to conventional T cells, NKT cell development requires interaction with membrane lymphotoxin-expressing cells, and lymphotoxin deficiency affected all three populations (CD4⁺, CD8⁺ and DN) of NKT cells. Their development is also absolutely dependent on pre-T-alpha signaling and at least partly dependent on granulocyte-macrophage colony-stimulating factor (GM-CSF) signaling. Like conventional T cells, peripheral NKT cells are regenerated from bone morrow in a relatively thymus-independent manner and their frequency in the peripheral blood and liver is about 0.1-0.5%. NKT cells require TCR-specific recognition in order to be activated, and their ligand is a conventional MHC-peptide complex. Most NKT cell seems to recognize CD1d in conjunction with hydrophobic ligands, although the precise nature of these ligands is not clear yet.[34]

The function most characteristic of CD4⁺ and DN thymic NKT cells is the rapid production of relatively high levels of immunoregulatory cytokines IL-4, IFN-gamma and TNF as well as potent lytic activity. They constitutively express Fas-L and can kill Fas⁺ target cells, including DN thymocytes; they can kill also tumor cells in a perforin-dependent manner. The role of NKT cells in immune responses is diverse.[33,34] They can control immune responses to infection and some tumors by: (a) driving the differentiation of Th2 responses mediated by their IL-4 production; (b) suppression of Th1 immune response through the production of IL-4, IL-10 and TGF-beta; (c) essential role in controlling anterior chamber-associated immune-deviation (ACAID), believed to prevent the eye from damage by inflammatory immune response; (d) the bone marrow DN NKT cells are important in preventing graft-versus-host disease following allogeneic bone-marrow transplantation in an IL-4-dependent way.

Myeloid Descendents

A common myeloid precursor gives rise to polymorphonuclear leukocytes (neutrophils, eosinophils, basophils), monocytes, platelets, and erythrocytes. Almost all myelo-monocytic cells (immature and mature) are positive for CD13 and CD33 antigens.[10] The myeloid cell-associated CD markers are summarized in Table 6 (for additional details see ref. 8).

Monocytes

Monocytes, members of the human mononuclear phagocyte system, are important in nonspecific defense against pathogens and in tumor surveillance. They also exert immunoregulatory functions via accessory cell activities and cytokine production. Monocytes derive from CD34+ precursors that express macrophage colony stimulating factor receptor (M-CSF-R) and/or Fcγ-receptor-I (CD64) and give rise exclusively to myelomonocytic cells. Neither of these receptors is found on early erythroid or lymphoid precursors. Immature myeloid monocytic precursors in bone marrow or cord blood are also characterized by high expression of CD135.[35] In peripheral blood, both CD34+ progenitor cells and mature monocytes may undergo further differentiation depending on the local environment. During tissue damage and infection, different components of pathogens, cytokines, chemokines, etc., are produced and deliver activation signals for the recruitment, activation, and differentiation of various leukocytes to the inflammatory site. Peripheral blood monocytes receiving such a stimulus enter the inflamed tissue and differentiate into macrophages, a process associated with functional alterations.[35] All monocytes/macrophages express the monocytic marker CD14, which is part of the lipopolysaccharide (LPS) receptor complex that is associated with the recently characterized toll-like receptor-4 (TLR4). Other members of the TLR family are also expressed on monocytes and myeloid cells and mediate pattern recognition signals.[36,37]

At least two different monocyte subsets can be distinguished based on phenotype and function.[38] A major subset of CD64 (FcγRI)+, CD14high monocytes exhibit high phagocytic activity, high production of inflammatory cytokines and reactive oxygen species, high ADCC and suppressor activity for antigen-activated lymphocytes. The small subset is represented by CD64-, CD14dim and loosely adherent HLA-DR+ cells with high IFNα producing capacity and exhibiting potent antigen presentation and T cell costimulatory capacity (Table 7). Phenotypic and functional heterogeneity has been described for tissue macrophages.

Dendritic Cells

Dendritic cells (DCs) are a distinct lineage of antigen-presenting leukocytes with potent capacity to induce primary T cell-mediated immune response.[39] DCs are characterized by distinct dendritic morphology, high expression of major histocompatibility complex (MHC) and costimulatory molecules, characteristic endocytic pathway for antigen uptake and processing, high capacity to stimulate resting T cells in an MHC-restricted fashion.[40] These attributes apply only to fully activated DC, whereas resting DCs and DC precursors lack some of them. In addition, DCs provide costimulatory signals required for T cells activation, secrete different cytokines (which can influence Th cells differentiation), enhance B lymphocyte expansion and antibody production, stimulate NK antibody-independent cytotoxic/cytolytic function and

Table 6. Myeloid/monocyte

Receptor	Family	Molecular Weight (kDa)	Distribution	Function
CD9 MRP-1	TM4-SF	24-26	Eosinophils, basophils, platelets, early B cells, activated T cells, endothelial cells, neurons, vascular smooth muscle, epithelia	Involved in cell adhesion and migration; role in platelet activation; ion channel/ transport molecule.
CD13 Amino-peptidase N	Metallo-protease	150	Granulocytes, monocytes and their precursors; endothelial cells, BM stromal cells, subset of large granular lymphocytes	Aminopeptodase N. Receptor for coronavirus; involved in interaction with cytomegalovirus
CD14 LPS-R	LRG (leucin-rich repeats)	55	Monocytes and granulocytes	Membrane bound; high-affinity receptor for LPS-LBL complexes on granulocytes, monocytes and macrophages. Soluble: binds LPS; required for LPS-induced activation of endothelial cells.
CD16 FcγRIII (note)	Ig-SF	50-65	NK cells, granulocytes, macrophages, subset of T cells	Low-affinity IgG receptor; main receptor for ADCC; activation of cytotoxicity, cytokine production and receptor expression.
CD16b FcγRIIIB	Ig-SF	50-65	Granulocytes	Low-affinity IgG receptor
CD23 FcεRII; Blast-2	C-type lectin	45	FcεRIIb: Macrophages, monocytes, eosinophils; FcεRIIa: activated B cells	Low-affinity Fc receptor for IgE; triggering of monokine release by human monocytes
CD32 FcγRII	Ig-SF	40	Granulocytes, macrophages, monocytes, eosinophils, basophils, B cells, platelets, Langerhans cells, some endothelium	Low-affinity receptor for IgG; mediates endocytosis, activation of secretion, cytotoxicity, immunomodulation
CD34	Sialomucin	90	Peripheral and mesenteric LN HEV. Haematopoietic stem/precursor cells. Widespread distribution on vascular endothelium	Mediates L-selectin-independent binding of leukocytes to HEV and L-selectin-independent binding of haematopoietic stem cells to stromal elements of bone marrow.

Table continued on next page

Table 6. **Continued**

Receptor	Family	Molecular Weight (kDa)	Distribution	Function
CD35	RCA	190-250	Neutrophils, monocytes, eosinophils, B cells, subset of T cells, erythrocytes, glomerular podocytes, FDCs, and some astrocytes	Receptor for C3b or C4b. Mediates adherence of C4b/C3b coated particles prior to phagocytosis. Major role in removing and processing immune complexes and facilitating their localization to lymphoid follicles.
CD36	Unassigned	88	Monocytes, macrophages, platelets, microvascular endothelial cells, DCs, mammary endothelial cells	Scavenger receptor recognition and phagocytosis of apoptotic cells; cell adhesion molecule in platelet-monocyte and platelet-tumor cell interactions
CD40	TNFR-SF	48	Macrophages, CD34$^+$ hematopoietic progenitors, B cells	Promotes cytokine production in macrophages and DCs
CD47	Ig-SF	50	Hematopoietic cells, endothelial cells, fibroblasts, epithelial tumor cells	Facilitates integrin-dependent adhesion
CD53	TM4-SF	32-42	All leukocytes	Signal transduction
CD54 ICAM-1	Ig-SF	90-115	Endothelial cells, DCs, epithelial cells, monocytes, B cells, T cells (low), fibroblasts	Ligand for LFA-1 and Mac-1. Mediates leukocyte adhesion to endothelium in inflammation: mediates T cell interactions with APCs or target cells and other T-T or T-B interactions
CD55 DAF	RCA	70	All leukocytes	Binds C3b and C4b to inhibit complement activation and deposition on plasma membranes. Binds C3bBb (alternative pathway convertase) and C4b2a (classical pathway convertase) to accelerate decay of the 3 convertase.

Table continued on next page

Table 6. Continued

Receptor	Family	Molecular Weight (kDa)	Distribution	Function
CD64 FcγRI	Ig-SF	72	Monocytes, macrophages, blood DCs, activated neutrophils, early myeloid lineage cells	High-affinity receptor for monomeric IgG promoting phagocytosis, ADCC, super-oxide and cytokine release in macrophage activation
CD66a	Ig-SF, CEA	140-180	Granulocytes	Plays signaling role and regulates adhesion activity of β2 integrin in neutrophils
CD69	C-type lectin (group V)	28/32	All activated leukocytes including neutrophils and eosinophils	Signal-transmitting receptor that may be involved in early events of monocyte activation
CD71 transferrin receptor	Unassigned	95	Macrophages and proliferating cells	Receptor for transferring; important in iron metabolism and cell growth; expressed by proliferating cells
CD74	Ig-SF	35-53	Macrophages, B cells, activated T cells	Chaperone promoting intracellular sorting of MHC class II molecules
CD88 C5aR	G-protein-coupled	43	Granulocytes, monocytes, DCs	C5a receptor
CD89 FcαR	Ig-SF	45-70	Neutrophils, monocytes, myeloid progenitors, activated eosinophils	Receptor for monomeric and polymeric forms of IgA and IgA2
CD90 Thy-1	IgSF	25-35	Hematopoietic stem cells	Contribute to the inhibition of proliferation/differentiation of hematopoietic stem cells
CD105 Endoglin	TGF-β receptor, type III	90	Activated monocytes and tissue macrophages	Receptor for TGF-β1 and β3; modulator of cellular responses to TGF-β1
CD114	Class I CKR-SF	130	Granulocytes, monocytes, granulocyte precursors	Receptor for G-CSF
CD115	Subclass II receptor tyrosine kinase	150	Monocytes, macrophages, and their progenitors	Receptor for M-CSF (CSF-1) which supports differentiation, survival and proliferation of mononuclear phagocytes

Table continued on next page

Table 6. Continued

Receptor	Family	Molecular Weight (kDa)	Distribution	Function
CD116	Class I CKR-SF	80	Macrophages, neutrophils, eosinophils and DCs	Binds GM-CSF with low affinity; binds GM-CSF with high affinity when co-expressed with CDw131
CD117	Receptor tyrosine kinase (type 3), Ig-SF	145-150	Hematopoietic stem and progenitor cells	Growth factor receptor which binds SCF, mast cell growth factor and Kit ligand
CDw119	Class II CKR-SF	95	Macrophages and most leukocytes	IFNγ-dependent biological signals
CD120a TNFRI, p55	TNFR-SF	55	Activated granulocytes and monocytes	Receptor for TNF; mediates anti-tumor effect; induces inflammatory response; upregulates leukocyte adhesion molecules and MHC class I
CD120b TNFRII, p75	TNFR-SF	75	Monocytes and macrophages	Receptor for TNF
CD121a	Ig-SF	75-85	All leukocytes	Binds biologically active IL-1 α and β, and the biologically inactive IL1ra
CDw128 IL-8R types A and B	α chemo-kines receptor	44-70	Neutrophils, monocytes, NK cells	IL-8 receptor
CD135 PLT3	Receptor tyrosine kinase	130	Multipotential myelomonocytic and primitive B cell progenitors	Growth factor receptor for early hematopoietic progenitors
CD164 MGC-24	Unassigned	80	Monocytes, bone marrow stromal cells	Thought to mediate adhesion between bone marrow stromal cells and hematopoietic progenitor cells
CD166 ALCAM	Ig-SF	100-105	Activated monocytes, activated T cells	Adhesion molecule

participate in homeostasis of immune system.[41] DCs do not express lineage the specific markers CD3, CD16 and CD19.

DCs can arise both from lymphoid and myeloid precursors (Table 8). Two types of myeloid DCs (interstitial DC and Langerhans-cell derived DC) and one type of lymphoid

Table 7. Phenotypic characteristics of human blood monocyte subsets

	Monocyte		Subsets	
Synonym names	"intermediate Mo" CD64⁻/CD16⁻ NIP	Fcγ-R-I (CD64)⁻ CD64⁻/CD16⁺ CD14^{dim}/CD16⁺	"regular Mo" CD64⁺/CD16⁺	Fcγ-R-I (CD64)⁺ CD64⁺/CD16⁻ CD14⁺/CD16⁻
Percentage within Mo	< 10%	< 10%	< 10%	< 80%
Phenotype				
Morphology	plasmacytoid	monocytic	monocytic	monocytic
Myeloid markers				
Esterase	-	+++	+++	+++
CD14	-	+	+++	+++
CD33	-	+	+++	+++
CD32 (FcγR-II)	-	+	+++	+++
CD68	+++	+++	+++	+++
Lymphoid markers				
CD3 (T cells)	-	-	-	-
CD19 (B cells)	-	-	-	-
CD56 (NK cells)	-	-	-	-
DC markers				
CD11c (integrin)	-	+++	+++	++
CD4	+++	+	+	+
CD123 (IL-3 receptor)	++	-	-	-
Other markers				
CD86 (B7-2)	-	+++	+++	++
MHC I-II	+	+++	+++	++
Function				
Cytokine release				
IFN-α	+++	+	+	+
IL-1, IL-6, TNF-α, CSF	-	+	+++	+++
PGE₂	-	-	+++	+++
IL-12	+	+++	+++	++
Phagocytosis & MPO activity	-	+	+++	+++
T/B-cell interaction				
APC	-	+++	+++	+
MLR	±	+++	++	±

(plasmocytoid) DCs have been identified.[41] The plasmocytoid DCs are localized in T cell zones of lymphoid organs or can exist as DC precursors in blood; they participate in CD4⁺ and CD8⁺ T cell priming and produce little or no IL-12. Myeloid DCs are also localized to T cell zones of organs and as DC precursors in peripheral blood, they can also occur as immature cells in epithelia and tissue interstitia. In contrast to plasmocytoid DCs, myeloid DCs produce large amounts of IL-12 and are superior in CD4⁺ and CD8⁺ cell priming. Some of the recently

Table 8. Human dendritic cells and markers

Postulated Lineage	Lymphoid *Plasmacytoid DC*	Myeloid *Interstitial DC*	*LC-Derived DC*
Blood precursors			
Phenotype	CD11c⁻ CD1a⁻ IL-3R⁺	CD11c⁺ CD1a⁺ IL-3R⁻	CD11c⁺ CD1a⁺ IL-3R⁻
IFN-α production	++++	-	-
Mature DCs			
Phenotype	CD11c⁻ IL-3R⁺ MHC class II⁺ CD11b⁻ CD13⁻ CD33⁻ CD4⁺⁺ CD1a⁻ Birbeck granule⁻ Langerin⁻ CD86⁺ CD40⁺ DC-LAMP⁺	CD11c⁺ IL-3R⁻ MHC class II⁺ CD11b⁺ CD13⁺ CD33⁺ CD4⁺ CD1a⁻ Birbeck granule⁻ Langerin⁻ CD86⁺ CD40⁺ DC-LAMP⁺	CD11c⁺ IL-3R⁻ MHC class II⁺ CD11b⁺ CD13⁺ CD33⁺ CD4⁺ CD1a⁻ Birbeck granule⁺ Langerin⁺ CD86⁺ CD40⁺ DC-LAMP⁺
Localization	T-cell zones of lymphoid organs; DC precursors in blood	T-cell zones of lymphoid organs; Germinal centers (GCDCs)?; DC precursors in blood; immature cells in tissue interstices (lungs, heart, kidney)	T-cell zones of lymph nodes; DC precursors in blood; immature cells in epithelia
Function			
IL-12 secretion	+/-	++++	++++
IL-10 secretion	-	++++	+/-
CD4⁺ T-cell priming	++	++++	++++
CD8⁺ T-cell priming	++	+++	++++
DC-B-cell interaction	?	++++	+

Abbreviations: DC= dendritic cell; IFN-α= interferon α; IL= interleukin; LC= Langerhans cell; MHC= major histocompatibility complex; R= receptor

identified dendritic cell-associated molecules, expressed during different stages of maturation include DC-LAMP, DC-CK1, Dectin-1 and-2, DC-SIGN, DORA, and TARK.[42]

International Nomenclature and Functional Characterization of Immunologic Markers

More than 166 different CDs were well characterized and approved for use in immunologic diagnoses during the last years. Hematopoietic differentiation schemes are based upon knowledge concerning normal hematopoiesis as well as the development of neoplastic cells in leukemias and nonHodgkin lymphomas (NHLs). Immunologic marker analysis can be applied for characterization of hematopoietic cells populations in normal, healthy populations, as well as in different autoimmune and primary or secondary immunodeficient diseases. The immune marker information is indispensable today for diagnosis and treatment of hematologic and solid tumors. With the expanding research in the area of signal transduction pathways and gene activation, highly specific molecular markers are emerging to further assist clinicians to distinguish normal from diseased cells and to design therapies targeted to the specific molecular defects characteristic of a particular disease.

The International workshops on Leukemia Marker Conference Analysis were started in 1981 and were held in every two years to update the list of CD molecules with nearly identified members. The following websites provide additional details about CD classifications:

http://www.ncbi.nlm.nih.gov/PROW/guide/45277084.htm

http://rncc.bidmc.harvard.edu/labs/viral/CD_Antigens_Table.html

http://www.researchd.com/rdicdabs/cdindex.htm

http://130.189.200.66/AboutFlow/cd_table.html

http://www.bioscience.org/atlases/cdclass/cdclass.htm

http://www.keratin.com/am/am025.shtml

http://www.univie.ac.at/Immunologie/cdguide.htm

References

1. Weissman IL. Translating stem and progenitor cell biology to the clinic: Barriers and opportunities. Science 2000; 287:1442-46.
2. Deans RJ, Moseley AB. Mesenchymal stem cells: Biology and potential clinical uses. Exp Hematol 2000; 28:875-84.
3. Van Dongen JJM, Comans-Bitter WM. Phenotypic and genotypic characteristics of the human prothymocyte. Immunol Res 1987; 6:250-62.
4. Janossy G, Tidman N, Papageorgiu ES et al. Distribution of T lymphocyte subsets in the human bone marrow and thymus: an analysis with monoclonal antibodies. J Immunol 1981; 126:1608-13.
5. Hockland P, Ritz J, Schlossman SF et al. Orderly expression of B cell antigens during the in vitro differentiation of nonmalignant human pre-B cells. J Immunol 1985; 135:1746-51.
6. Reinherz EL, Haynes BF, Nadler LM et al, eds. Leucocyte typing II. Volume 1: Human myeloid and Hematopoietic cells. Berlin: Springer-Verlag, 1986.
7. Knapp W, Dörken B, Rieber EP et al, eds. Leucocyte Typing IV. White cell differentiation antigens. Oxford: Oxford University Press, 1989.
8. Immune Receptor Supplement. 2nd ed. Immunol Today 1997.
9. Akashi K, Reya T, Dalma-Weiszhausz D et al. Lymphoid precursors. Curr Opin Immunol 2000; 12:144-150.
10. Griffin JD, Ritz J, Nadler LM et al. Expression of myeloid differentiation antigens on normal and malignant myeloid cells. J Clin Invest 1981; 68:932-41.
11. Grabbe S, Kämpgen E, Schuler G. Dendritic cells: multi-lineal and multi-functional. Trends Immunol Today 2000; 21:431-433.
12. Ghia P, ten Boekel E, Rolink AG et al. B-cell development: a comparison between mouse and man. Rev Immunol Today 1998; 19:480-485.
13. Hardy RR, Hayakawa K. B Cell Development Pathways. Annu Rev Immunol 2001; 19:595-621.
14. Martin F, Kearney JF. B1 cells: Similarities and differences with other B cell subsets. Curr Opin Immunol 2001; 13:195-201.
15. Fagarasan S, Honjo T. T-Independent Immune Response: New Aspects of B Cell Biology. Science 2000; 290:89-92.
16. Tsubata T. Co-receptors on B lymphocytes. Curr Opin Immunol 1999; 11:249-255.
17. Berg LJ, Kang J. Molecular determinants of TCR expression and selection. Curr Opin Immunol 2001; 13:232-241.
18. Asnagli H, Murphy KM. Stability and commitment in T helper cell development. Curr Opin Immunol 2001; 13:242-247.
19. Malek TR, Porter BO, He Y-W. Multiple γc-dependent cytokines regulate T-cell development. Viewpoint Immunol Today 1999; 20:71-76.
20. Steele CR, Oppenheim DE, Hayday AC. $\gamma\delta$ T cells: Nonclassical ligands for nonclassical cells. Current Biol 2000; 10:R282-R285.

21. Hamann D, Roos MThL, van Lier RAW. Faces and phases of human CD8⁺ T-cell development. Viewpoint Immunol Today 1999; 20:177-180.
22. Constant SL, Bottomly K. Induction Of Th1 and Th2 CD4⁺ T-Cell responses: The alternative approaches. Annu Rev Immunol 1997; 15:297-322.
23. den Haan JMM, Bevan MJ. A novel helper role for CD4 T cells. PNAS 2000; 97:12950-12952.
24. Abbas AK, Murphy KM, Sher A. Functional diversity of helper T lymphocytes. Nature 1996; 383:787-793.
25. Mosmann TR, Li L, Sad S. Functions of CD8 T-cell subsets secreting different cytokine patterns. Semin Immunol 1997; 9:87-92.
26. Watts TH, DeBenedette MA. T cell co-stimulatory molecules other than CD28. Curr Opin Immunol 1999; 11:286-293.
27. Mueller DL. T cells: A proliferation of costimulatory molecules. Curr Biol 2000; 10:R227-R230.
28. Raulet DH. Development and tolerance of natural killer cells. Curr Opin Immunol 1999; 11:129-134.
29. Liu C-C, Perussia B, Young JD-E. The emerging role of IL-15 in NK-cell development. Trends Immunol Today 2000; 21(3)113-116.
30. Waldmann TA, Tagaya Y. The multifaceted regulation of interleukin-15 expression and the role of this cytokine in NK cell differentiation and host response to intracellular pathogens. Annu Rev Immunol 1999; 17:19-49.
31. Cooper MA, Fehniger TA, Turner SC et al. Human natural killer cells: a unique innate immunoregulatory role for the CD56^bright subset. Blood 2001; 97:3146-3151.
32. Yokoyama WM. Natural killer cell receptors. Curr Opin Immunol 1998; 10:298-305.
33. Elewaut D, Kronenberg M. Molecular biology of NK T cell specificity and development. Semin Immunol 2000; 12:561-568.
34. Hammond KJL, Pelikan SB, Crowe NY et al. NKT cells are phenotypically and functionally diverse. Eur J Immunol 1999; 29:3768-3781.
35. Valledor AF, Borràs FE, Cullell-Young M et al. Transcription factors that regulate monocyte/macrophage differentiation. J Leukocyte Biol 1998; 63:405-417.
36. Poltorak A, He X, Smirnova I et al. Defective LPS Signaling in C3H/HeJ and C57BL/10ScCr Mice: Mutations in Tlr4 Gene. Science 1998; 282(5396):2085-2088.
37. Medzhitov R, Janeway C Jr. Advances in Immunology: Innate Immunity. N Engl J Med 2000; 343(5):338-344.
38. Phenotypic and Functional Properties of Human Blood Monocyte Subsets. J Leukocyte Biol 2001; 69:14 (Table 1)
39. Reid SD, Penna G, Adorini L. The control of T cell responses by dendritic cell subsets. Curr Opin Immunol 2000; 12:114-121.
40. Banchereau J, Steinman RM. Dendritic cells and the control of immunity. Nature 1998; 392:245-252.
41. Hartgers FC, Figdor CG, Adema GJ. Towards a molecular understanding of dendritic cell immunobiology. Trends Immunol Today 2000; 21:542-545.
42. Pulendran B, Maraskovsky E, Banchereau J et al. Modulating the immune response with dendritic cells and their growth factors. Trends Immunol 2001; 22:41-47.

Additional References for CD Markers

CD1
1. Calabi F, Jarvis JM, Martin L et al. Two classes of CD1 gene. Eur J Immunol 1989; 19:285-292.
2. Zeng Z, Castano AR, Segelke BW et al. Crystal structure of mouse CD1: an MHC-like fold with a large hydrophobic binding groove. Science 1997; 277:339-345.
CD2
1. Davis SJ, van der Merwe PA. The structure and ligand interactions of CD2: implications for T-cell function. Immunol Today 1996; 17(4):177-87.
2. Davis SJ, Ikemizu S, Wild MK et al. CD2 and the nature of protein interactions mediating cell-cell recognition. Immunol Rev 1998; 163:217-36.

CD3
1. Alcover A, Alarcon B. Internalization and intracellular fate of TCR-CD3 complexes. Crit Rev Immunol 2000; 20(4):325-46
2. Gobel TW, Bolliger L. Evolution of the T cell receptor signal transduction units. Curr Top Microbiol Immunol 2000; 248:303-20.

CD4
1. Li S, Satoh T, Korngold R et al. CD4 dimerization and oligomerization: Implications for T-cell function and structure-based drug design. Immunol Today 1998; 19(10):455-62.
2. Surman S, Lockey TD, Slobod KS et al. Localization of CD4+ T cell epitope hotspots to exposed strands of HIV envelope glycoprotein suggests structural influences on antigen processing. Proc Natl Acad Sci USA 2001; 98(8):4587-92.

CD5
1. Youinou P, Jamin C, Lydyard PM. CD5 expression in human B-cell populations. Immunol Today 1999; 20(7):312-326.
2. Pers JO, Jamin C, Predine-Hug F et al. The role of CD5-expressing B cells in health and disease. Int J Mol Med 1999; 3(3):239-45.

CD7
1. Sempowski GD, Lee DM, Kaufman RE et al. Structure and function of the CD7 molecule. Crit Rev Immunol 1999; 19(4):331-48.
2. Tien HF, Wang CH. CD7 positive hematopoietic progenitors and acute myeloid leukemia and other minimally differentiated leukemia. Leuk Lymphoma 1998; 31(1-2):93-8.

CD8
1. Leahy DJ. A structural view of CD4 and CD8. FASEB J 1995;9(1):17-25.
2. Gao GF, Jakobsen BK. Molecular interactions of coreceptor CD8 and MHC class I: the molecular basis for functional coordination with the T-cell receptor. Immunol Today 2000; 21(12):630-6.

CD9
1. Clay D, Rubinstein E, Mishal Z et al. CD9 and megakaryocyte differentiation. Blood 2001; 97(7):1982-9.
2. Rubinstein E, Benoit P, Billard M et al. Organization of the human CD9 gene. Genomics 1993; 16(1):132-8.
3. Boucheix C, Benoit P, Frachet P et al. Molecular cloning of the CD9 antigen. A new family of cell surface proteins. J Biol Chem 1991; 266(1):117-22.

CD10
1. McCormack RT, Nelson RD, LeBien TW. Structure/function studies of the common acute lymphoblastic leukemia antigen (CALLA/CD10) expressed on human neutrophils. J Immunol 1986; 137(3):1075-82.
2. Cutrona G, Ferrarini M. Expression of CD10 by human T cells that undergo apoptosis both in vitro and in vivo. Blood 2001; 97(8):2528.
3. Uherova P, Ross CW, Schnitzer B et al. The clinical significance of CD10 antigen expression in diffuse large B-cell lymphoma. Am J Clin Pathol 2001; 115(4):582-8.

CD13
1. Wex Th, Bühling F, Arndt M et al. The activation-dependent induction of APN-(CD13) in T-cells is controlled at different levels of gene expression. FEBS Letters 1997; 412:1:53-56.
2. Riemann D, Kehlen A, Langner J. CD13—not just a marker in leukemia typing. Immunol Today 1999; 20(2):83-8.

CD14
1. Landmann R, Muller B, Zimmerli W. CD14, new aspects of ligand and signal diversity. Microbes Infect 2000; 2(3):295-304.
2. Gregory CD. CD14-dependent clearance of apoptotic cells: relevance to the immune system. Curr Opin Immunol 2000; 12(1):27-34.

CD16
1. Kato K, Fridman WH, Arata Y et al. A conformational change in the Fc precludes the binding of two Fcgamma receptor molecules to one IgG. Immunol Today 2000; 21(7):310-2.
2. Ravetch JV, Lanier LL. Immune inhibitory receptors. Science 2000; 6;290(5489):84-9.

CD19

1. Fujimoto M, Poe JC, Hasegawa M et al. CD19 regulates intrinsic B lymphocyte signal transduction and activation through a novel mechanism of processive amplification. Immunol Res 2001; 22(2-3): 81-298.

2. Rolink AG, Schaniel C, Busslinger M et al. Fidelity and infidelity in commitment to B-lymphocyte lineage development. Immunol Rev 2000; 175:104-11.

3. Fearon DT, Carroll MC. Regulation of B lymphocyte responses to foreign and self-antigens by the CD19/CD21 complex. Annu Rev Immunol 2000; 18:393-422.

CD20

1. Polyak MJ, Tailor SH, Deans JP. Identification of a cytoplasmic region of CD20 required for its redistribution to a detergent-insoluble membrane compartment. J Immunol 1998; 161(7):3242-8.

2. Feuring-Buske M, Buske C, Unterhalt M et al. Recent advances in antigen-targeted therapy in nonHodgkin's lymphoma. Ann Hematol 2000; 79(4):167-74.

CD21

1. Zabel MD, Weis JH. Cell-specific regulation of the CD21 gene. Int Immunopharmacol 2001; 1(3):483-93.

2. Chen Z, Koralov SB, Kelsoe G. Regulation of humoral immune responses by CD21/CD35. Immunol Rev 2000; 176:194-204.

3. Reid RR, Woodcock S, Prodeus AP et al. The role of complement receptors CD21/CD35 in positive selection of B-1 cells. Curr Top Microbiol Immunol 2000; 252:57-65.

CD22

1. Nitschke L, Floyd H, Crocker PR. New functions for the sialic acid-binding adhesion molecule CD22, a member of the growing family of Siglecs. Sc and J Immunol 2001; 53(3):227-34.

2. Smith KG, Fearon DT. Receptor modulators of B-cell receptor signaling—CD19/CD22. Curr Top Microbiol Immunol 2000; 245(1):195-212.

3. Tsubata T. Co-receptors on B lymphocytes. Curr Opin Immunol 1999; 11(3):249-55.

CD23

1. Riffo-Vasquez Y, Pitchford S, Spina D. Murine models of inflammation: role of CD23. Allergy 2000;55 Suppl 61: 21-6.

2. Heyman B. Regulation of antibody responses via antibodies, complement, and Fc receptors. Annu Rev Immunol 2000; 18:709-37.

3. Fremeaux-Bacchi V, Kolb JP, Rakotobe S et al. Functional properties of soluble CD21. Immunopharmacology 1999; 42(1-3):31-7.

CD24

1. Watts TH, DeBenedette MA. T cell co-stimulatory molecules other than CD28. Curr Opin Immunol 1999; 11(3):286-293.

2. Ellis TM, Moser MT, Le PT et al. Alterations in peripheral B cells and B cell progenitors following androgen ablation in mice. Int Immunol 2001;13(4):553-8.

3. Suzuki T, Kiyokawa N, Taguchi T et al. CD24 induces apoptosis in human B cells via the glycolipid-enriched membrane domains/rafts-mediated signaling system. J Immunol 2001; 166(9):5567-77.

CD25

1. Shakib F, Schulz O, Sewell H. A mite subversive: cleavage of CD23 and CD25 by Der p 1 enhances allergenicity. Immunol Today 1998; 19(7):313-6.

2. Shevach EM. Certified Professionals. 4CD4(+)CD25(+) suppressor T cells. J Exp Med 2001; 193(11):F41-6.

3. Wolf M, Schimpl A, Hunig T. Control of T cell hyperactivation in IL-2-deficient mice by CD4(+)CD25(-) and CD4(+)CD25(+) T cells: Evidence for two distinct regulatory mechanisms. Eur J Immunol 2001; 31(6):1637-45.

CD26

1. Van Damme J, Struyf S, Wuyts A et al. The role of CD26/DPP IV in chemokine processing. Chem Immunol 1999; 72:42-56.

2. Dong RP, Tachibana K, Hegen M et al. Determination of adenosine deaminase binding domain on CD26 and its immunoregulatory effect on T cell activation. J Immunol 1997; 159(12):6070-6.

CD28
1. Holdorf AD, Kanagawa O, Shaw AS. CD28 and T cell co-stimulation. Rev Immunogenet 2000; 2(2):175-84.
2. Salazar-Fontana LI, Bierer BE. T-lymphocyte coactivator molecules. Curr Opin Hematol 2001; 8(1):5-11.

CD30
1. Muta H, Boise LH, Fang L et al. CD30 signals integrate expression of cytotoxic effector molecules, lymphocyte trafficking signals, and signals for proliferation and apoptosis. J Immunol 2000; 165(9):5105-11.
2. Stein H, Foss HD, Durkop H et al. CD30(+) anaplastic large cell lymphoma: A review of its histopathologic, genetic, and clinical features. Blood 2000; 96(12):3681-95.

CD32
1. Sinclair NR. Fc-signalling in the modulation of immune responses by passive antibody. Scand J Immunol 2001; 53(4):322-30.
2. Lyden TW, Robinson JM, Tridandapani S et al. The Fc receptor for IgG expressed in the villus endothelium of human placenta is Fc gamma RIIb2. J Immunol 2001; 166(6):3882-9.

CD35
1. Carroll MC. CD21/CD35 in B cell activation. Semin Immunol 1998; 10(4):279-86.
2. Eggleton P, Tenner AJ, Reid KB. C1q receptors. Clin Exp Immunol 2000; 120(3):406-12.

CD36
1. Abumrad N, Coburn C, Ibrahimi A. Membrane proteins implicated in long-chain fatty acid uptake by mammalian cells: CD36, FATP and FABPm. Biochim Biophys Acta 1999; 1441(1):4-13.
2. Nicholson AC, Febbraio M, Han J et al. CD36 in atherosclerosis. The role of a class B macrophage scavenger receptor. Ann NY Acad Sci 2000; 902:128-31.

CD37
1. Knobeloch KP, Wright MD, Ochsenbein AF et al. Targeted inactivation of the tetraspanin CD37 impairs T-cell-dependent B-cell response under suboptimal costimulatory conditions. Mol Cell Biol 2000; 20(15):5363-9.
2. Maecker HT, Todd SC, Levy S. The tetraspanin superfamily: molecular facilitators. FASEB J 1997; 11(6):428-42.

CD38
1. Lund FE, Muller-Steffner HM, Yu N et al. CD38 signaling in B lymphocytes is controlled by its ectodomain but occurs independently of enzymatically generated ADP-ribose or cyclic ADP-ribose. J Immunol 1999; 162(5):2693-702.
2. Bergthorsdottir S, Gallagher A, Jainandunsing S et al. Signals that initiate somatic hypermutation of B cells in vitro. J Immunol 2001; 166(4):2228-34

CD40
1. Schattner EJ. CD40 ligand in pathogenesis and therapy. Leuk Lymphoma 2000; 37(5-6):461-72.
2. Diehl L, Den Boer AT, van der Voort EI et al. The role of CD40 in peripheral T cell tolerance and immunity. J Mol Med 2000; 78(7):363-71.
3. Hurwitz AA, Kwon ED, van Elsas A. Costimulatory wars: the tumor menace. Curr Opin Immunol 2000; 12(5):589-96.

CD40-CD40 ligand
van Kooten C, Banchereau J. CD40-CD40 ligand. J Leukocyte Biol 2000; 67:2-17.

CD43
1. van den Berg TK, Nath D, Ziltener HJ et al. Cutting edge: CD43 functions as a T cell counterreceptor for the macrophage adhesion receptor sialoadhesin (Siglec-1). J Immunol 2001; 166(6):3637-40.
2. Corinti S, Fanales-Belasio E, Albanesi C et al. Cross-linking of membrane CD43 mediates dendritic cell maturation. J Immunol 1999; 162(11):6331-6.

CD45
1. Penninger JM, Irie-Sasaki J, Sasaki T et al. CD45: New jobs for an old acquaintance. Nat Immunol 2001; 2(5):389-96.
2. Alexander DR. The CD45 tyrosine phosphatase: A positive and negative regulator of immune cell function. Semin Immunol 2000; 12(4):349-59.

3. Ashwell JD, D'Oro U. CD45 and Src-family kinases: And now for something completely different. Immunol Today 1999; 20(9):412-6.

CD47

1. Brown EJ, Frazier WA. Integrin-associated protein (CD47) and its ligands. Trends Cell Biol 2001; 11(3):130-5.
2. Pettersen RD. CD47 and death signaling in the immune system. Apoptosis 2000; 5(4):299-306.

CD53

1. Beinert T, Munzing S, Possinger K et al. Increased expression of the tetraspanins CD53 and CD63 on apoptotic human neutrophils. J Leukoc Biol 2000; 67(3):369-73.
2. Engering A, Pieters J. Association of distinct tetraspanins with MHC class II molecules at different subcellular locations in human immature dendritic cells. Int Immunol 2001; 13(2):127-34.

CD55

1. Miwa T, Song WC. Membrane complement regulatory proteins: insight from animal studies and relevance to human diseases. Int Immunopharmacol 2001; 1(3):445-59.
2. Smith GL. Vaccinia virus immune evasion. Immunol Lett 1999; 65(1-2):55-9.

CD56

1. Pittet MJ, Speiser DE, Valmori D et al. Cutting edge: Cytolytic effector function in human circulating CD8+ T cells closely correlates with CD56 surface expression. J Immunol 2000; 164(3):1148-52.
2. Moretta L, Biassoni R, Bottino C et al. Human NK-cell receptors. Immunol Today 2000; 1(9):420-22

CD64

1. Rouard H, Tamasdan S, Moncuit J et al. Fc receptors as targets for immunotherapy. Int Rev Immunol 1997; 16(1-2):147-85.
2. Tamm A, Schmidt RE. IgG binding sites on human Fc gamma receptors. Int Rev Immunol 1997; 16(1-2):57-85.

CD66

1. Billker O, Popp A, Gray-Owen SD et al. The structural basis of CEACAM-receptor targeting by neisserial Opa proteins. Trends Microbiol 2000; 8(6):258-60.

CD69

1. Marzio R, Mauel J, Betz-Corradin S. CD69 and regulation of the immune function. Immunopharmacol Immunotoxicol 1999; 21(3):565-82.
2. Solana R, Mariani E. NK and NK/T cells in human senescence. Vaccine 2000; 18(16):1613-20.

CD72

1. Parnes JR, Pan C. CD72, a negative regulator of B-cell responsiveness. Immunol Rev 2000; 176:75-85.
2. Glynne R, Ghandour G, Rayner J et al. B-lymphocyte quiescence, tolerance and activation as viewed by global gene expression profiling on microarrays. Immunol Rev 2000; 176:216-46.

CD74

1. Shih L, Ong GL, Burton J et al. Localization of an antibody to CD74 (MHC class II invariant chain) to human B cell lymphoma xenografts in nude mice. Cancer Immunol Immunother 2000; 49(4-5):208-16.

CD79

1. Engel P, Tedder TF. New CD from the B cell section of the Fifth International Workshop on Human Leukocyte Differentiation Antigens. Leuk Lymphoma 1994; 13 (S1):61-4.
2. Paloczi K, Batai A, Gopcsa L et al. Immunophenotypic characterisation of cord blood B-lymphocytes. Bone Marrow Transplant 1998; 22 (S4):89-91.
3. Koyama M, Ishihara K, Karasuyama H et al. T CD79 alpha/CD79 beta heterodimers are expressed on pro-B cell surfaces without associated mu heavy chain. Int Immunol 1997; 9(11):1767-72.

CD80

1. Shortman K, Wu L. Parentage and heritage of dendritic cells. Blood 2001; 97(11):3325.
2. Balbo P, Silvestri M, Rossi GA et al. Differential role of CD80 and CD86 on alveolar macrophages in the presentation of allergen to T lymphocytes in asthma. Clin Exp Allergy 2001; 31(4):625-36.

CD81
1. Levy S, Todd SC, Maecker HT. CD81 (TAPA-1): a molecule involved in signal transduction and cell adhesion in the immune system. Annu Rev Immunol 1998; 16:89-109.
2. Rice CM. Is CD81 the key to hepatitis C virus entry? Hepatology 1999; 29(3):990-2.

CD83
1. Scholler N, Hayden-Ledbetter M, Hellstrom KE et al. CD83 is a sialic acid-binding Ig-like lectin (Siglec) adhesion receptor that binds monocytes and a subset of activated CD8+ T cells. J Immunol 2001; 166(6):3865-72.
2. Cramer SO, Trumpfheller C, Mehlhoop U et al. Activation-induced expression of murine CD83 on T cells and identification of a specific CD83 ligand on murine B cells. Int Immunol 2000; 12(9):1347-51.

CD86
1. Kobata T, Azuma M, Yagita H et al. Role of costimulatory molecules in autoimmunity. Rev Immunogenet 2000; 2(1):74-80.
2. Bechmann I, Peter S, Beyer M et al. Presence of B7--2 (CD86) and lack of B7--1 (CD(80) on myelin phagocytosing MHC-II-positive rat microglia is associated with nondestructive immunity in vivo. FASEB J 2001;15(6):1086-8.

CD88
1. Meddows-Taylor S, Pendle S, Tiemessen CT. Altered expression of CD88 and associated impairment of complement 5a-induced neutrophil responses in human immunodeficiency virus type 1-infected patients with and without pulmonary tuberculosis. J Infect Dis 2001; 183(4):662-5.
2. Crass T, Ames RS, Sarau HM et al. Chimeric receptors of the human C3a receptor and C5a receptor (CD88). J Biol Chem 1999; 274(13):8367-70.

CD89
1. Bracke M, Lammers JW, Coffer PJ et al. Cytokine-induced inside-out activation of FcalphaR (CD89) is mediated by a single serine residue (S263) in the intracellular domain of the receptor. Blood 2001; 97(11):3478-83.
2. Honorio-Franca AC, Launay P, Carneiro-Sampaio MM et al. Colostral neutrophils express Fc alpha receptors (CD89) lacking gamma chain association and mediate noninflammatory properties of secretory IgA. J Leukoc Biol 2001; 69(2):289-96.

CD90
1. Thornley I, Sutherland DR, Nayar R et al. Replicative stress after allogeneic bone marrow transplantation: changes in cycling of CD34+CD90+ and CD34+CD90- hematopoietic progenitors. Blood 2001; 97(6):1876-8.
2. Wuchter C. Impact of CD133 (AC133) and CD90 expression analysis for acute leukemia immunophenotyping. Haematologica 2001; 86(2):154-61.

CD95
1. Catlett IM, Xie P, Hostager BS et al. Signaling through MHC class II molecules blocks CD95-induced apoptosis. J Immunol 2001; 166(10):6019-24.
2. De Maria R, Testi R. Fas-FasL interactions: A common pathogenetic mechanism in organ-specific autoimmunity. Immunol Today 1998; 19(3):121-5.

CD100
1. Hall KT, Boumsell L, Schultze JL et al. Human. CD100, a novel leukocyte semaphorin that promotes B-cell aggregation and differentiation. Proc Natl Acad Sci USA 1996; 93(21):11780-5.
2. Elhabazi A, Delaire S, Bensussan A et al. Biological activity of soluble CD100. I. The extracellular region of CD100 is released from the surface of T lymphocytes by regulated proteolysis. J Immunol 2001; 166(7):4341-7.

CD105
1. Li C, Hampson IN, Hampson L et al. CD105 antagonizes the inhibitory signaling of transforming growth factor beta1 on human vascular endothelial cells. FASEB J 2000; 14(1):55-64.
2. Kumar S, Ghellal A, Li C et al. Breast carcinoma: Vascular density determined using CD105 antibody correlates with tumor prognosis. Cancer Res 1999; 59(4):856-61.

CD114

1. Jilma B, Hergovich N, Homoncik M et al. Granulocyte colony-stimulating factor (G-CSF) downregulates its receptor (CD114) on neutrophils and induces gelatinase B release in humans. Br J Haematol 2000; 111(1):314-20.

2. Hollenstein U, Homoncik M, Stohlawetz PJ et al. Endotoxin down-modulates granulocyte colony-stimulating factor receptor (CD114) on human neutrophils. J Infect Dis 2000; 182(1):343-6.

CD116

1. Dabusti M, Castagnari B, Moretti S et al. CD116 (granulocyte-macrophage colony stimulating factor receptor). J Biol Regul Homeost Agents 2001; 15(1):86-9.

CD122

1. Corrigall VM, Arastu M, Khan S et al. Functional IL-2 receptor beta (CD122) and gamma (CD132) chains are expressed by fibroblast-like synoviocytes: Activation by IL-2 stimulates monocyte chemoattractant protein-1 production. J Immunol 2001; 166(6):4141-7.

2. Goldrath AW, Bogatzki LY, Bevan MJ. Naïve. T cells transiently acquire a memory-like phenotype during homeostasis-driven proliferation. J Exp Med 2000; 192(4):557-64.

CD124

1. Koubek K, Kumberova A, Stary J et al. Expression of cytokine receptors on different myeloid leukemic cells. Neoplasma 1998; 45(4):198-203.

CD125

1. Yamada K, Yamakawa M, Imai Y et al. Expression of cytokine receptors on follicular dendritic cells. Blood 1997;90(12):4832-41.

CD126

1. Muller-Newen G, Kuster A, Hemmann U et al. Soluble IL-6 receptor potentiates the antagonistic activity of soluble gp130 on IL-6 responses. J Immunol 1998;161(11):6347-55.

CD 132

1. Nielsen OH, Kirman I, Johnson K et al. The circulating common gamma chain (CD132) in inflammatory bowel disease. Am J Gastroenterol 1998; 93(3):323-8.

CD135

1. Bertho JM, Chapel A, Loilleux S et al. CD135 (Flk2/Flt3) expression by human thymocytes delineates a possible role of FLT3-ligand in T-cell precursor proliferation and differentiation. Scand J Immunol 2000; 52(1):53-61.

2. Rappold I, Ziegler BL, Kohler I et al. Functional and phenotypic characterization of cord blood and bone marrow subsets expressing FLT3 (CD135) receptor tyrosine kinase. Blood 1997; 90(1):111-25.

CD154

1. Fiumara P, Younes A. CD40 ligand (CD154) and tumour necrosis factor-related apoptosis inducing ligand (Apo-2L) in haematological malignancies. Br J Haematol 2001; 113(2):265-74.

2. Skov S, Bonyhadi M, Odum N et al. IL-2 and IL-15 regulate CD154 expression on activated CD4 T cells. J Immunol 2000; 164(7):3500-5.

CD164

1. Watt SM, Chan JY. CD164—a novel sialomucin on CD34+ cells. Leuk Lymphoma 2000; 37(1-2):1-25.

2. Zannettino AC, Buhring HJ, Niutta S et al. The sialomucin CD164 (MGC-24v) is an adhesive glycoprotein expressed by human hematopoietic progenitors and bone marrow stromal cells that serves as a potent negative regulator of hematopoiesis. Blood 1998; 92(8):2613-28.

CD166

1. van Kempen LC, Nelissen JM, Degen WG et al. Molecular basis for the homophilic ALCAM-ALCAM interaction. J Biol Chem 2001; 276(28):25783-90.

2. Bowen MA, Aruffo AA, Bajorath J. Cell surface receptors and their ligands: in vitro analysis of CD6-CD166 interactions. Proteins 2000; 40(3):420-8.

CHAPTER 2

Application of Monoclonal Antibodies to the Diagnosis and Classification of Acute Leukemias

Nóra Regéczy

More than Routine Immunophenotyping of Acute Leukemias

The classification of acute leukemias (AL) should reflect as objectively as possible the biology and the clinical features of the many different disorders presenting as proliferations of hematopoietic precursors or blast cells (Fig. 1). The traditionally used method of classifications divides AL into morphologically distinct subtypes. Over past decades, there have been many advances in the understanding of genetic factors in the biology of these neoplasms, particularly the acute leukemias.

The newly proposed WHO classification of hematological malignancies stratifies the disorders according to a combination of morphology, immunophenotype, genetic features and clinical syndromes. The WHO defined beyond the lineage assignment, specific subclassification of the tumors for understanding the prognosis and selection of more specific therapeutic approaches, wich disease entities can be recognized by pathologist and have clinical relevance. Although, flow cytometry has evolved more than an indispensable tool to hematologic diagnosis, the multiparameter immunophenotyping are based on gating and separating techniques to isolate a most homogenous blast population combined morphological, immunological and special genetic data. Thus, we will summarize each of major subtypes of AL incorporating the morphologic, immunophenotypic and cytogenetic, molecular approach in the delineation of biologically important subgroups to support a combined modality attitude to the daily practice of hemato-oncology.

The FAB and the New WHO Classification

The French, American, British (FAB) classification introduced in the late 1970s and refined ten years later with M7 has been the basis for most studies to date.[1,2] After the FAB group defined relatively precise criteria for the morphological differentiation of acute lymphoid leukemia (ALL) from acute myeloid leukemia (AML) and for the identification of major subtypes, the further studies failed to show the high interobserver reproducibility for the clinically important distiction of ALL from AML (range of 95% or higher), in the case of subtyping distinct smaller entities the diagnostic concurrence was only 60%-80%.[3,4] The second major point of criticism of the FAB morphological-cytochemical classification relates to its apparent lack of prognostic significance. A series of large studies failed to show a prognostic relationship

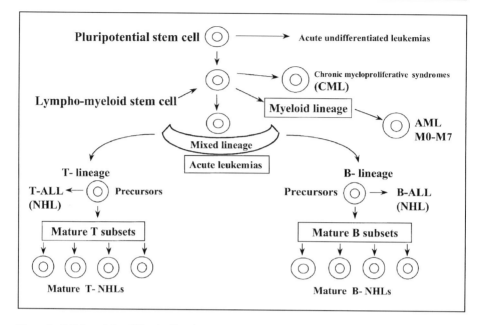

Figure 1. Cellular origin of blood cell malignancies.

between the different myeloid FAB subtypes.[4] Regarding the lymphoid FAB subtypes (L1-L3) it is generally agreed that L3 morphology conveys a bad prognosis.

Advances in the fields of immunology (monoclonal antibodies), cytogenetics (chromosome abnormalities) and molecular genetics (gene rearrangement, protooncogens) have introduced a greater complexity with some important evidence of prognostic utility which recently needed to be taken into consideration by WHO in order to improve both the diagnostic precision and the reproducibility of the classification.[5-12]

The WHO adopts not only the clinical features such as prior therapy and history of myelodysplasia, but also the previous description of morphology, immunology and cytogenetics (MIC) proposals which emerged from the knowledge that some ALs are probably better defined by chromosome translocation and that immunophenotyping is an essential component in the investigation of acute leukemias.[7-12]

Immunophenotyping

The multidimensional flow cytometry using hundreds of monoclonal antibodies (MoAbs), improved gating strategies, such as coupled analysis of common leucocyte antigen CD45 with SSC-granularity and quantitation of antigen expression has become an important diagnostic tool in the differential diagnosis of acute leukemias.[13,14] Multivariate analysis has permitted a powerful determination of the composite phenotypic pattern of each individual tumor and documented that immunophenotypes are independent prognostic factors. Therefore, in light of its high reproducibility and prognostic significance, immunophenotyping could merge several diagnostic processes achieving an essential role in the diagnosis and categorization of acute leukemias.[6-12,15-21]

The previous hypothesis for classification systems has considered that the leukemic cell has a normal counterpart due to ALs derive from immature lympho-hematopoietic cells with the definitive phenotypic features of these precursors that are also found in the corresponding malignancy.[22] Oppositely, the current knowledge has confirmed that the relationship of blasts with normal precursors is more contoversial.[22,24] Since all antigens are expressed normally during hematopoietic development, it is worth to know that the staining is not simply related to the contaminating normal cells included in the tumor. Excluding the normal cells from the analysis is esssential on the basic knowledge of antigen expression on both normal and abnormal cells.[19,22-26] The phenotypic abnormalities presented in leukemias could be categorized into four major groups, but not rarely more than one abnormality expressed: (a) lineage infidelity, (b) maturational asynchrony, (c) antigenic absence, (d) quantitation abnormalities.[24]

a. Because of the neoplastic cells were not representative of a frozen state of normal development but were truly aberrant, only the particular constellation of antigens, the complex patterns should be the basis for comparing leukemic cells to the normal cells, especially the lineage infidelity appears. Various terms have been used to designate these unusual acute leukemias showing both myeloid and/or lymphoid features: hybrid, biphenotypic, biclonal, bilineal, chimaeric, mixed or infidelity lineage and multiphenotypic.[22-25]

b. Some leukemias expressed maturational asynchony such as the appearance of early precursor antigens (CD34, terminal deoxynucleotidyl transferase: Tdt) with mature markers (CD11b, CD20 or CD3).[15,17-19,22-26]

c. A common example of antigenic absence is in the case of some ALL without CD45 panleukocyte antigen expression.[15-19,27,28]

d. The leukemia is different not only from normal but also from every other leukemia due to dysregulation in tumorigenesis proceed in maturational stages. These uncoordinated regulation results multivariant differences in phenotype and antigenic intensities expressed by the various hematological neoplasm.[19,23,24,27,28]

Basic Distinction of AML versus ALL

Traditionally, the major point is to distinguish myeloid from lymphoid leukemia that is cardinal in the view of therapeutic and prognostic importance. In this regard, immunological methods are the basis, but there is an evidence that no single marker identifies every case of leukemias. Beyond lineage assignment, the more homogenous grouping of cases, specific subclassification is essential for defining the clinically and biologically relevant subtypes of ALL and AML. Although no real consensus exists, the basic panel of immunological markers—more than 30 antigens—used to distinguish properly AML from ALL, as well as T-lineage ALL from B- lineage ALL is shown in Table 1.

The most specific antigens for defining lineage are expressed in the cytoplasm (intracytoplasmatic: cy): myeloperoxidase (MPO), cyCD79a and cyCD3 for myeloid, B- and T-lineages respectively, and the cyμ, cyCD22 are also provided analysis the more mature stages of B cell differentiation (Table 2).[11,15,16,25,26,29] Moreover, the detection of surface and/or intracytoplasmatic antigens are also sensitive for the classification of rare subtypes as megakaryocytic (CD41, CD42, CD61) and erythroid (glycophorin A) markers.[11] The leukemic cells in all myelocytic and monocytic subtypes express a various combinations of CD13, CD33, CD65, CD117 and MPO.[11,15,16,30,31]

On the cell surface, the CD13 is an important pan myeloid marker which is, however, not absolutely specific, as about 10-20% of ALL cases are CD13 positive, as well.[15,18,24,31-33] The T cell antigen (Ag) CD7 is found in all stages of T-ALL and the B cell Ag CD19 is characteristic of all stages of B-lineage ALL.[11,18,19] Although CD7 expression has been also reported on about 10-30% of AML cases.[24,34-38] The combinations of surface membrane (sm) CD13 with cytoplasmic MPO staining, the smCD7 with cyCD3 as well as smCD19 with cyCD22 are definitive indicators of AML and T or B lineage ALL, respectively.[19,39,40] The vast majority of cases (> 90%) can be assigned confidently to the major classes of acute leukemia on the basis of these marker profiles.

Table 1. Panel of markers for the diagnosis of acute leukemias

	Monoclonal Antibodies	Specificity
1st panel	cyCD22, cyCD79a, CD19, CD10	B lineage
	cyCD3, CD7, CD2,	T lineage
	cyMPO, CD13, CD33, CD65, CD117*	Myeloid
	Tdt, CD34, HLA-DR	Nonlineage
2nd panel	cyμ, smIg,κ,λ,CD20,CD24	B lymphoid
	CD1a, CD3, CD5, CD4, CD8, TCR	T lymphoid
	CD41, CD42, CD61	Megakaryocytic
	glycophorin A	Erythroid
Optional	CD14, CD15, CD11, cy-lysozime, CD36, CD38, CD68	Myeloid

cy= cytoplasmic; sm= surface membrane; * known as stem cell factor or c-kit, too (References: 11, 15, 16, 19, 20, 23, 26, 29-31)

Table 2. Basic monoclonal antibody panel for the distinction of ALL versus AML

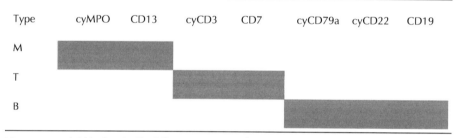

Type	cyMPO	CD13	cyCD3	CD7	cyCD79a	cyCD22	CD19
M							
T							
B							

M= acute myeloid leukemia; B= B-lineage ALL; T= T-lineage ALL; cy= intracytoplasmic

Immunological Classification of ALL

The WHO did not retain the FAB classification of ALL: L1, L2 and L3 which does not correlate well with the immunologic classification and the genetic abnormalities; only L3 morphology seems to be restricted to the B-ALL.[12,18,41] As to date, the genetic as well as immunophenotypic markers should be important prognostic factors and allow more adequate therapeutic approaches.

The most immature lymphoid leukemia is characterized by stem cell markers showing HLA-DR, CD34 and terminal deoxynucleotidyl transferase (Tdt) positivity (Figs. 2, 3, 4, 5).[11,18] The enzyme Tdt is expressed on the nuclear membrane of immature leukemic lymphoid cells, but is absent in more mature lymphoid leukemias. This expression pattern is related to the important role of Tdt in the rearrangement of immunoglobulin (Ig) and T cell receptor (TCR) genes during the early stages of lymphopoiesis.[29] Both Tdt and CD34 are restricted to immature lymphoid leukemias. Although, the diagnostic specificity of cyTdT is limited by its presence 20% in cases of AML, but the level of expression is higher in case of ALL than AML.[11,35-38,42]

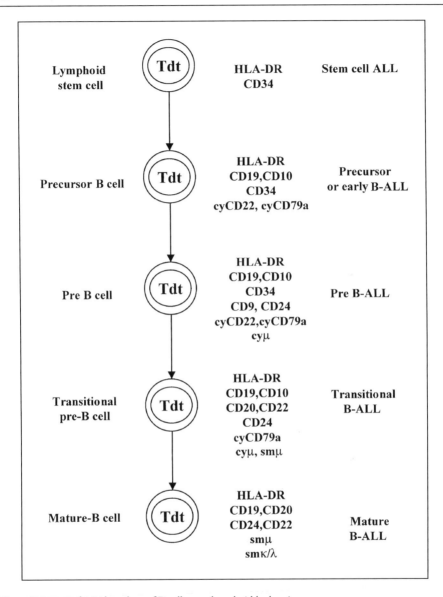

Figure 2. Immunological markers of B-cell acute lymphoid leukemias.

Immunophenotypic Profile of the B-Lineage ALL

The aquisition of knowledge in the field of leukemia immunophenotyping continues to be a dynamic process. The phenotyping plays a central role in the determination of clinically relevant subsets of B-ALL. The proposed classification should be based on the pattern of reactivity to a panel of antibodies rather than any specific reagent. We divide ALL into precursor B, pre-B, transitional pre-B and B-ALL, and T-ALL (Table 3 and Fig. 2)[15,18,31]

The most immature cell of B-lineage ALL (precursor B cell) can be characterized by Tdt, CD34, HLA-DR, CD19, cyCD79a with variable cyCD22, CD10 and CD20. (Figs. 2, 3). One early and highly specific intracytoplasmatic marker is the CD79a besides cyCD22.[43] Some

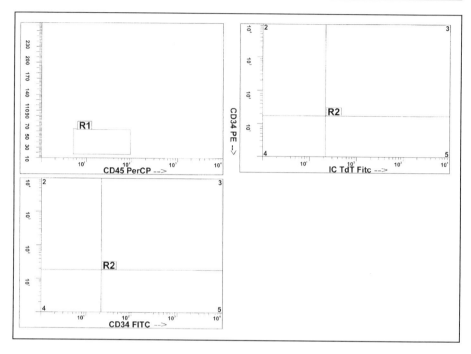

Figure 3. Acute lymphoblastic leukemia with cyCD79a, Tdt and CD34 positivity: precursor B-ALL.

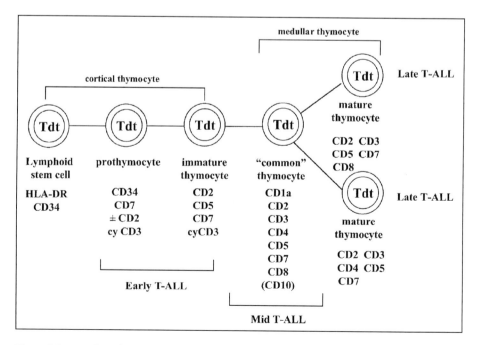

Figure 4. Immunological markers of acute lymphoid leukemias with T-cell type.

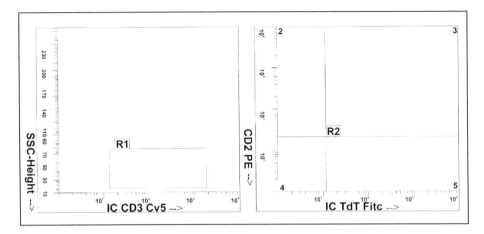

Figure 5. Acute lymphoblastic leukemia with cyCD3, Tdt and CD2 positivity: early T-ALL.

cases may have very low or absent CD45.[27,28] Recently, ALL may be classified on the basis of DNA content measured also by flow cytometry. Moreover, the relatively absence of the pan-leukocyte antigen (CD45) and the higher DNA index of blasts was correlated with well respond to therapy and long time survival.[18,28]

In another, earlier proposed classification the presence of CD10(CALLA) considered as an entity (common B-ALL), but the CD10 antigen may be expressed in all types of B lineage ALL.[11] Among early B-ALL cases, CALLA is associated with excellent prognosis and a clearly discriminated subtype harbors frequently hyperploid DNA content and t(12;21) creates TEL/ AML1 product.[15,31,44] Oppositely, the absence of CD10 correlated with the aberrant expression of myeloid antigens particularly CD15, and these phenotypes are typically presented with translocations involving the MLL gene at t(4;11) or 11q23 and (9;22) rather with poor prognosis.[15,18,31,32,41,45] The incidence of myeloid antigen positive ALL is approximately 5-25% or higher.[15,18,41] However, most of the lineage disrupted leukemias: My+ALL (myeloid antigen positive ALL) or Ly+AML and the biphenotypic acute leukemias correlates with these cytogenetic abnormalities.[15,18,31,32,35,36,46]

Cells of the more mature pre-B ALL express CD19, HLA-DR, cyCD79a, cyCD22, cyμ and CD10. Tdt and CD20 are variable, but CD34 generally negative. A most common and

---

Table 3. Classification of acute lymphoid leukemia with B cell type

B-ALL Subgroups   Antigens

	HLA-DR	CD34	Tdt	CD22	cyCD79a	CD19	CD20	CD10	Igμ	κ/λ
Precursor	+	+	+	cy+	+	+	-	+	-	-
Pre	+	+	+	+	+	+	-	+	cy+	-
Transitional	+	-	-+	+	+	+	+	+-	+	-
Mature	+	-	-+	+	+	+	+	+-	+	+

cy= intracytoplasmic; Tdt= terminal deoxynucleotidyl transferase specific subtype has t(1;19) and it is characterized by worse outcome and by the absence of CD34 and the presence of CD10 and CD9.[15,31,47] The absence of CD34 seems to be an independent poor prognostic factor in general in B-lineage ALL.[48]

The transitional pre-B cell ALL has been newly designated by expression of both cytoplasmic and surface immunoglobulin heavy chains without light chains. The phenotype usually is as in pre-B ALL with more favorable prognosis.[18,31,49]

As the result of the WHO classification, ALL presented a leukemic phase of any precursor lymphoid neoplasm. However, the B-ALL is quite similar to the Burkitt's lymphoma in leukemic phase having t(8;14).[12,18,41] On the CD45-SSC plot, this leukemia display larger size than other ALL. The phenotype shows the B-lineage antigens (CD19, CD20, CD22) and HLA-DR with bright clonal smIg including either cytoplasmic or surface ligth chain expression. The absence of immature markers, CD34 and Tdt is characteristics. The defining characteristics is the surface Ig, also in many cases with CD10 positivity or distinct morphology.[18] Rarely cases expressing light chain are seen without L3 morphology.[15,31] Conversely, occasional cases may present similar phenotype to precursor B-ALL (smIg negative) not only with L3 morphology and cytoplasmatic ligth chain expression but also with t(8;14) that should be considered functionally equivalent to B-ALL.[15,50]

Immunophenotypes of T Cell ALL

T-ALL have commonly larger lymphoblastic scattergram without measurable granularity. The most sensitive marker of T-ALL is CD7, but questionable specificity in the view of CD7 or other lymphoid marker positive AML cases.[18,19,34-38] The detection of particularly sensitive and specific cyCD3 must strongly confirm the T lineage assignment (Table 4 and Figs. 4, 5).[11,29]

The most immature differentiation stage of T-ALL (also designated as pre-T or early T-ALL) is characterized by CD7 positivity and intracytoplasmic CD3: features which appear simultaneously or more often precede the expression of the CD2 antigen (Fig. 5).

The next differentiation stage of T-ALL (mid or common) is characterized by especially CD1a and double CD4 and CD8 with pan T antigens (CD7, CD2, CD5) which are common thymocyte markers, respectively. In these cases Tdt is frequently positive with minimal or absent surface CD3.

The cells of the mature T cell leukemias (medullary, mature or late) express pan T markers (CD7, CD2, CD5) and segregated CD4 or CD8 with CD3. It is important to notice that leukemias with double negative cells (CD3+, CD4-,CD8-) represent a distinc leukemic cell population.

Table 4.   *Classification of acute lymphoid leukemia with T cell type*

T-ALL subgroups

	CD7	CD2	CD3	CD5	CD1a	CD4	CD8
Early	+	+-	cy+	+-	-	-	-
Mid	+	+	+	+	+	+	+
Mature	+	+	+	+	-	+/-	+/-

+/-= No simultaneous expression of CD4 and CD8; cy= intracytoplasmic

The common phenotypic feature of T-ALL is the abnormal selection from pan T cell antigens that results the loss of these specific antigens or aberrant combinations.[18,19,31] Although the adult pre-T ALL has been shown to differ significantly in clinical features and prognosis from more differentiated phenotypic subtypes the prognostically important T-ALL subtypes are not well defined.[15,18,41,45] Only the one suggested prognostic factor should be the CD2. Thus, patients have CD2 positive T blasts may have also higher survival rate than others with CD2 negative T-ALL.[51] Moreover, the CD10 negative T-ALL cases appear to do worse.[15,18,53]

Acute Lymphoid Leukemias Expressing Myeloid-Associated Antigens

In 30 to 40% of adult ALL the lymphoblasts express antigens characteristic of myeloid cells.[18,24,32,33] There are very few antigens considered reliable in defining strictly "myeloid" versus "lymphoid" antigens. Furthermore, the stringency of the criteria used to diagnose My+ALL has varied but finally in 1995 has concluded.[11] However, immunophenotyping and genetic studies have led to recognition of lineage disrupted and true biphenotypic ALs associated with specific genetic alterations as a potentially important prognostic factors, such as t(9;22) and MLL gene rearrangements t(4;11) or (11q23).[15,18,31,32,35,36,46] Expression of myeloid antigens in adult ALL is clearly associated with a lower likelihood of achieving remission.[18,45]

Immunophenotypic Classification of Acute Myeloblastic Leukemia

Immunophenotyping allows an early and rapid diagnosis of AL and established the lineage assignment. The classification of AML, from M1 to M7, is still largely based on morphology and cytochemistry.[1,2] Studies with MoAbs also provide beyond the phenotype a special cellular morphology on scattergram using CD45 versus SSC and could show effectively the different cell populations, demonstrates each major lineage in the bone marrow. Multiparameter analysis is a powerful tool to detect some very large variables such as size, shape and granularity during myeloid maturation pathway thus we summarized AML phenotypes below in the context of FAB categories (Table 5 and Figs. 6, 7).[6,13,14,19]

However, the newly described WHO classification summarized the myeloproliferative disorders correlating not only with MIC abut also with clinical behavior and genetic markers. Within AML four main groups are recognized: (I) AML with recurrent cytogenetic translocations, (II) AML with myelodysplasia-related features, (III) Therapy-related AML and MDS, IV. AML not otherwise specified, respectively.[12]

The major changes that have had an impact on flow cytometric analysis are concluded in this classification: above 20% blast count should define AML, cytogenetic aberrations are recongnized as distinct categories as well as dysplasia and therapy are included in AML diagnosis.

In the case of AML, the valuable markers, beyond cyMPO, CD13, CD33, CD65 and CD117 (known as stem cell factor or c-kit, too) myeloid antigens are correlated closely with differentiation features.[6,11,16,19,30] These features could be e.g., the lack of HLA-DR expression on promyelocytes of M3 and the bright expression of the CD14 antigen in leukemias with a monocytic component (M4 and M5), megakaryoblastic leukemia (M7) by MoAbs (CD41, CD42, CD61) and erythroleukemia (M6) with glycophorin A, respectively.[11,16,19]

Immunophenotyping is also of prognostic importance in AML. The immunological definition of AML are based on two or more myeloid antigen positivity, however all of the cyMPO, CD13, CD33, CDw65 and CD117 antigen expressed phenotypic pattern seems to be a homogenous subgroup in AML predicting higher complete remission rate and overall survival.[30] The expression of TdT, CD7, CD2 or CD19 in a given percentage of blast cells may influence the biologic features of AML.[24,34-38] Oppositely with ALL, CD34 positivity of blast cells adversely affects remission induction and duration of survival after intensive chemotherapy independently from secondary AML and pre-existing myelodysplasia.[34,36,40,53]

Table 5.   The classification of acute myeloblastic leukemias

FAB Type	AML Subgroups	Myeloid Markers Expressed	Nonlineage Markers Expressed
M0	Early myeloblastic leukemia	cyMPO-+, CD13, CD33, CD117+-*	CD34, HLA-DR, Tdt+-
M1	Undifferentiated myeloblastic leukemia	cyMPO, CD13, CD33, CD11b-+, CD15-+, CD65-+, CD117+-	CD34+-, HLA-DR, Tdt-+
M2	Differentiated myeloblastic leukemia	cyMPO, CD13, CD33, CD15, CD65-+, CD117	CD34+-, HLA-DR, Tdt-+
M3	Promyelocytic leukemia	cyMPO, CD13-+, CD33, CD15	
M4	Myelomonocytic leukemia	cyMPO+-, CD13-+, CD33, CD14+-, CD11b+-, CD65+-, CD117	CD34+-, HLA-DR+-
M5a	Monoblastic leukemia	cyMPO-+, CD13-+, CD33, CD14+-, CD11b+-, CD65, CD117+-	CD34-+, HLA-DR+-
M5b	Differentiated monocytic leukemia	cyMPO-+, CD13-+, CD33, CD14, CD11b, CD65, CD117-+	HLA-DR
M6	Erythroblastic leukemia	cyMPO-+, CD13+-, CD33+-, CD65, CD117, Gly A, H Ag	CD34, HLA-DR
M7	Megakaryoblastic leukemia	CD13, CD33, CD117, CD41, CD42, CD61	CD34, HLA-DR

cy= intracytoplasmic; MPO= myeloperoxidase; GlyA= glycophorin A; * known as stem cell factor or c-kit too; +-= variable often positive; -+= variable often negative

Early Myeloblastic Leukemia: AML-M0

A distinct group of cases, 3% of all AML have blasts with immature morphology (with low forward scatter as lymphoblasts often resembling ALL L2) and negative cytochemical tests which are insufficient for them to be classified as AML by FAB criteria.[6,11,16,54] B and T specific lymphoid markers are negative, except for the Tdt which is expressed in half of the cases.[36,37,54]

The key finding to define these cases as AML-M0 is their reactivity with one MoAbs against myeloid associated antigens, usually CD13, CD33, CD117 (known as stem cell factor or c-kit, too) or cyMPO.[11,30,54-58] M0 blasts are often positive for HLA-DR and CD34. This group represents an early form of AML which can be distinguished from ALL (positive lymphoid markers) and other types of AML (positive cytochemistry).

Confirmation of the myeloblastic nature of AML-M0 can be obtained by means of ultrastructural cytochemistry with the demonstration of small MPO positive granules.[56] In the absence of MPO, leukemias should be classified as M0 only also in the absence of other lineage restricted antigens.

Our experience in acute leukemias using the MoAb anti-MPO, which detects the intracytoplasmic proenzyme form and the alpha-chain of the MPO molecule, has confirmed the valuable sensitivity and specificity of this reagent when used by flow cytometry.[57] However, the ultrastructural demonstration of MPO is slightly more sensitive than the MoAb anti-MPO for the detection this enzyme in very early myeloid cells.[56] The expressed lymphoid markers on M0 blasts, such as CD7 or CD4, and the CD34 are in association with worse prognosis. Among the patients with AML M0, frequently could be detected complex cytogenetic abnormalities involving chromosomes 5 or 7 and the presence of lymphoid markers correlated with t(9;22) or MLL gene aberrations.[15,55,58]

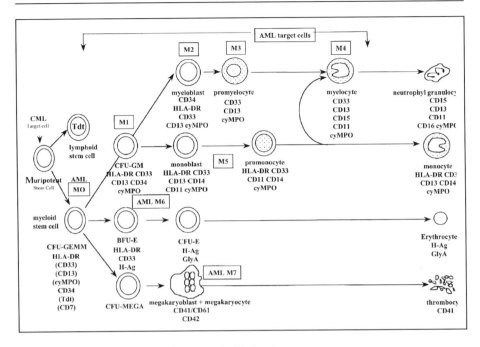

Figure 6. Immunological markers of acute myeloid leukemias.

AML M1-M2

During the maturation the flow appearance of M1 and M2 demonstrates a continuum from blasts changed rather with granularity (higher SSC). Moreover, at least one prominent expression of myeloid antigens: brighter CD13 and CD15 than CD33, particularly on M2 blasts with HLA-DR, but reduced CD34 may be detected. The blast count should be lower, but cyMPO positive blast are the predominant population in the bone marrow (Fig. 7).[15,40,56] The presence of lymphoid antigens as CD19 and less often CD56 is correlated with translocation involved (8;21) and favorable prognosis in adults.[34,35,53,56]

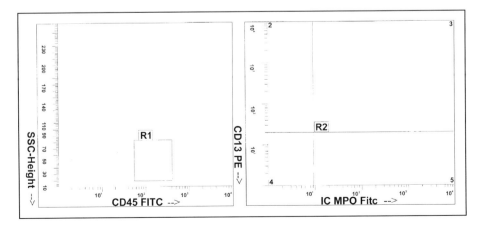

Figure 7. Acute myeloid leukemia (AML M1) with cyMPO and CD13 positive blast population.

AML M3

This morphologically exclusive type of AML shows these features also on CD45-SSC plot. Large size and high granularity, lower level of CD45 similarly to myeloblasts, except a micro-, hypogranular form of this type (M3v).[13,19] The characterization of M3 cells should be based on these morphology and high autofluorescence, the lack of surface HLA-DR with dense expression of CD33 and less prominent other pan myeloid antigens (CD13, CD11b and CD15) and CD34.[15,31,36,60,61] In spite of this association, this phenotype still form a heterogenous group of patients, due to this limitation immunophenotyping of AML M3 has been considered as a secondary diagnostic tool almost exclusively restricted to supporting the morphologic diagnosis.[40]

The presence of CD2 lymphoid marker on both M3 and M3v blast associated with translocation (15;17) rather than t(11;17) and favorable prognosis due to retinoic acid treatment.[34,36,58,60-63]

There are newly described, rarely cases of myeloid/NK AML with CD56 positivity and similar morphology to M3 tend to be unexpectedly refractory to this therapy.[64]

AML M4 and M5

Typically these monocytoid forms of AML demostrate an abnormal overlapping scattergram from myelo-monoblasts to mature monocytic region making it difficult to gate on either population exclusively.

The phenotype requires the pan myeloid antigen expressions CD33, CD13, HLA-DR, during maturation variable CD11b and CD14. The lack of CD34 and more dense CD33 positivity characterized the M5, particularly the more matured M5b cells.[15,31] These forms of AML rarely presented lymphoid antigens, such as CD4, CD2 or CD19.[31,34,35,53] The important subtype of M4 is M4Eo has the CD2 positive phenotype pattern frequently involved the alterations of 16 chromosome with marrow eosinophilia.[15,65]

An other rare but distinct form of AML is characterized by monocytic features, erythrophagocytosis and bleeding diathesis with coagulation features suggestive of fibrinolysis and not of disseminated intravascular coagulation (DIC) as seen in M3. The chromosome abnormality t(8;16) (p11;p13) is a consistent finding in these patients.[6]

AML M6

This subtype of acute myeloid leukemia is rare and not well characterized with flow cytometry. The erythroid components of bone marrow show as a CD45 negative and variable CD71 as well as CD13, CD33 and HLA-DR, CD34 positive population. The most specific antigen, glycophorin A may be seen by intracytoplasmic staining.[16,19,66]

AML M7

It is quite difficult to detect reliable by known diagnostic methods in hematology this standalone form of AML. Thus flow cytomerty should be achieve an applicable assay for describing M7. Because of the platelets are often adhere to blast fals positive reactions with megakaryocyte-platelet antigens may be seen. The lineage of M7 blasts must be confirmed with CD33, CD34 or HLA-DR antigens rather by ultrastructural platelet peroxidase assay or by intracytoplasmic staining of CD42, CD41 or CD61.[67]

AML with Trilineage Myelodysplasia

An important minority of primary AML (between 10-15% of cases) show marked myelodysplastic features involving erythroid, granulocytic and megakaryocytic cells. Trilineage myelodysplasia could be seen in all FAB subtypes of AML except M3; it is more common in

M6 and M7 and rare in M1. It is presumable that these cases represent myelodysplastic syndrome (MDS) presenting as AML. No specific immunologic markers are known in this peculiar type of AML.[6,12]

Acute Myeloid Leukemia Expressing Lymphoid Markers

The incidence of lymphoid-associated antigen expression in AML (Ly+AML) has been measured extremely variable. The results range from 13% to 60%. The wide variability is related to the analysis of different study population (children versus adults), to the stringency of the initial definition of AML, and to various technical aspects such as use of flow cytometry, immunofluorescence microscopy, or immunoenzymatic staining.[22,24,34-38]

The strict definition of lymphoid marker positive myeloid leukemia is the expression of one or two lymphoid antigens without fulfilling the criteria for biphenotypic acute leukemia.[11,46] A wide range of lymphoid-associated markers are expressed on AML cells (CD1, CD2, CD3, CD4, CD5, CD7, CD8, CD10, CD19, CD20, CD21, CD22).[11,15,16,22,24,34-38,40,53-56,60-63] Most of these markers fall within the 0-10% range of positivity but CD7 and CD4 antigens are expressed on 15% and 24% of AML cells, respectively. According to the newly proposed criteria, the cut-off point to consider a marker positive at 20% of blast cells with exception for cytoplasmatic CD3, CD79a and Tdt.[11] The simultaneous expression of CD2 and CD7 in AML cases is suggested to be a nonrandom event being associated with distinct clinical and prognostic features.

Recent advances in understanding the biologic function of CD antigens has considered their lack of lineage fidelity. CD10 and CD13 are membrane associated enzimes, other surface antigens have originally been demonstrated to be expressed on both (normal) lymphoid and myeloid cells: CD4, present on T cells and monocytes; CD10, expressed on pre-B cells and granulocytes; due to its broad spectrum of reactivity CD7 is no longer considered by some investigators to be solely a T-lineage antigen, but it appears to identify also a particular subset of myeloid progenitors.[15,19]

Except for CD4, CD2, CD7 and CD56 no correlation was found between FAB morphology and the expression of any of the antigens studied e.g., CD10, CD19.[35,36] The frequent accurence of the CD4 antigen in M4 and M5 AML is not unexpected. CD7 was most often detected in subtype M0 and M1.[36,55-58] CD2 expression is frequently associated with M4Eo and M3 morphology.[36,65] It is of particular importance to distinguish cases of myeloid marker positive early T-ALL from CD7 positive AML. Such a distinction seems to be especially difficult in adult cases and might require analysis of cytoplasmic CD3 and myeloperoxidase.[11,29,57]

Biphenotypic Acute Leukemia

The most common problems leading to discrimination of true biphenotypic acute leukemia (BAL) cases, the features of samples, the failure to exclude normal cells from analysis, the type, number and lineage specifity of antigens studied (there is also a significant epitopic heterogeneity among monoclonal antibodies in a given CD group) are resolved by single or multi-color staining.[6,13,21,24,26,46]

Findings on the BAL, as defined by the scoring proposed support that such cases arise from early progenitors or stem cells exhibiting both myeloid and lymphoid features shown not only by immunological markers but the Ig and TCR gene configuration (Table 6).[68] When a proper definition for BAL or mixed-lineage AL is applied the score is greater than two from two separate lineages, excluding the My+ALL and Ly+AML cases.[11,46] Application of this scoring might be useful to distinguish the true biphenotypic leukemias and will help in the future to disclose whether biphenotypic leukemia represents a distinct clinicobiological entity. Thus, these previously unanticipated reactivities support the assumption that only few antigenic determinants are entirely restricted to a particular cell lineage the most important lineage-specific antigens are cytoplasmic CD22 and CD79a, CD3 and MPO, respectively.

Table 6.   *Scoring system for the definition of acute biphenotypic leukemia**

Score Points	B Lymphoid	T Lymphoid	Myeloid
2	cyCD79α, cyCD22, cyµ	cyCD3, TCR	cyMPO
1	CD10, CD19, CD20	CD2, CD5, CD8, CD10	CD13, CD33, CD65, CD117**
0.5	Tdt, CD24	Tdt, CD7	CD14, CD15, CD64

* A case is defined biphenotypic when scores greater than 2 points for the myeloid and one lymphoid lineage;[11,16,46,69] ** known as stem cell factor or c-kit too

Aberrant phenotype pattern of BAL correlate highly with specific genetic alterations, particularly MLL gene alterations and translocation of the (9;22) chromosomes.[6,15-17,23-25,31,46,69,70]

Undifferentiated Acute Leukemia

Rare subtypes of acute leukemia remain unclassifiable even after extensive analysis, these are recognised as acute undifferentiated leukemia (AUL).

The AUL blast cells do not express lineage specific markers. Although these cases are often CD34+, HLA-DR+, CD38+ and Tdt+, or other CD7, CD9 antigens without lineage commitment.[11,15,31]

Minimal Residual Disease in Acute Leukemia

The aim of treatment of AL is ultimately to eradicate the tumor cells. Recently, many approaches play a role in the assessment and monitoring of minimal residual disease (MRD) in the undetectable range by conventional morphology and cytochemistry.[71-79]

The possibility of early investigation of the real clonogenic cells is of great clinical interest and value not only to predict relapse but also the rate of contamination of leukemic cell in samples collected for autologous transplantation. In this point of view, the sensitivity of these methods is the most important question.

There are many advances of using flow cytometry in detection of MRD: higher sensitivity than traditional morphology and cytogenetic studies, high specificity and the more rapid, simple and reproducible data acquisition than also highly (most) sensitive and specific favorable molecular polymerase chain reaction (PCR) techniques.[15,18,19,31,71-74] At the time, the use of PCR for the analysis of specific gene rearrangements and the identification of phenotype patterns have emerged the most attractive ones. In spite of these advantages of flow cytometry, it is worth to know that major disadvantages also exist: there are no leukemia specific antigens and several phenotypic changes may be investigated during relapse.[19,31,34,75-77] If leukemia specific pattern is detected, the only question remained to solve whether abnormal phenotype is stable.

The leukemia associated phenotype generally results from linegae infidelity, asynchronous antigen expression, and antigen overexpression that are above reviewed.[19,24,34]

There are other situations to help investigating MRD by cytometry. Some phenotypic combinations found on leukemic blasts are restricted to certain normal tissues e.g., bone marrow or thymus[15,18,19,24,73-77] and ectopic phenotype may be demonstrated if e.g., normal cortical thymocytes are outside the thymus. Therefore, even a TdT positive and pan T antigen (CD7, CD2, CD5) positive cells in the peripheral blood, bone marrow or central nervous system of T-ALL patients indicates residual disease.[18,71,73,74] Immature cells with CD34 or Tdt positivity are seen in central nervous system could lead the diagnosis of leukemia. Second, in a higher proportion of B lineage and myeloid acute leukemias blasts express combinations of normal differentiation antigens which are not found or extremely rare among normal hematopoietic cells.[15,18,19,74-78] In B lineage ALL leukemia-associated phenotypes include Tdt, CD34 and CD19, a combination that is very rare or absent in normal B cell precursors. One of the most inconsistent result of immunophenotyping is in the case of higher proportion of reactive B cell precursors. After chemotherapy or transplantation, increased numbers of CD10 and CD19 positive B cell precursors are in the peripheral blood as well as bone marrow, and the percentage of these cells may suggest AL.[18,19,73-75] Without any tumor specific phenotype the maturational spectrum represents the one detectable feature of reactive, normal B cell precursors. In AML, leukemia-associated features include the expression of myeloid-associated antigens cyMPO, CD13, CD33, CD65 and CD117 (CD117 known as stem cell factor or c-kit, too) with CD7, CD19, CD56 and/or Tdt.[22,24,34-36,42,53,74,76-79] Third, the existence of light scatter changes demonstrated abnormal cells, such as more forward scatter in the case of myelo- or lymphoblasts.[75,76]

Although there is a little doubt about the reliability of immunologic techniques to identify small numbers of leukemic cells, at levels below 10^{-4}, their value relative to that of other methods for detecting minimal residual disease needs to be evaluated.[71-74,78] The general limitations of the immunophenotyping are dictated by the size of the sample required and by the uneven distribution of residual leukemia. Clinical observations have shown that bone marrow samples taken from different sites may contain different numbers of leukemic blasts. Nonetheless, all PCR methods also have limitations, including amplifications of genome parts from nonviable or nondividing leukemic cells.[71,74]

Advantages of immunologic techniques over other available methods include speed, as well as the ability to precisely quantitate the number of residual leukemic cells and to assess their viability. Moreover, the detection of neoplastic cells, whether by flow cytometry or molecular methods, at a single point posttreatment probably does not predict relapse as reliably as changing values over time. Thus, future technical efforts in this area should focus on the precise follow-up of patients with multicolor analysis selecting more homogenous blast populations and on the identification of new leukemia-associated phenotypes.

Multidrug-Resistance in Acute Leukemia

The prognostic implications of flow cytometry as are the follows: the investigation of leukemia types with variable risk and the monitoring of residual disease as well as to approximate the response to therapy with a newly described feature of several tumors, the multidrug-resistance.[15,16,30,36,80-83]

The identification of a phenotypically distinct subset of drug-resistant leukemic blasts may be provided by immunological and molecular methods, respectively. It has recently been shown, the expression and the function of drug transporters, such as MDR1 is associated with CD34 positivity and poorer outcome in AML.[15,30,80-82] The major clinical aspect of multidrug-resistance is how effectively the multidrug transporters function regarding the ability to response to therapy.[16,18,30,81-83] The functional activity of these proteins affected significantly the clinical outcome and the alternative treatment applicated drug-resistance modulating agents. The antigen detection or quantitation of resistance limited by the wide range of expressed drug-transporters are investigated regarding multidrug-resistance phenotype.[30,36,80] In spite of this limitation, there is a valuable funcional test using selective inhibitors for known transporters that allows a discrimination not only the resistant phenotype but also the related proteins separately. Thus, clinically relevant multidrug-resistance assay should be selected properly to assess the clinical relevance of further therapeutic trials.

References

1. Bennett JM, Catovsky D, Daniel MT et al. Proposed revised criteria for the classification of acute myeloid leukemia. Ann Intern Med 1985; 103:620-25.

2. Bennet JM, Catovsky D, Daniel MT et al. Criteria for the diagnosis of acute leukaemia of megakaryocyte lineage (M7). Ann Intern Med 1985; 103:460-62.

3. Head DR, Savage RA, Cerezo L et al. Reproducibility of the French-American-British classification of acute leukemia: the Southwest Oncology Group experience. Am J Hematol 1985; 18:47-57.

4. Drexler HG, Menon M, Klein M et al. Correlation of surface marker expression with morphologically and immunologically defined subclasses of acute myeloid leukaemias. Clin Exp Immunol 1986; 65:363-72.

5. Bain B, Catovsky D. Practical aspects of the classification of acute leukaemia. J Clin Pathol 1990; 43:822-87.

6. Catovsky D, Matutes E, Buccheri V et al. A classification of acute leukemia for 1990s. Ann Hematol 1991; 62:16-21.

7. First MIC Cooperative Study Group. Morphologic, immunologic and cytogenetic (MIC) working classification of acute lymphoblastic leukemias. Cancer Genet Cytogenet 1986; 23:189-97.

8. Second MIC Cooperative Study Group. Morphologic, immunologic and cytogenetic (MIC) working classification of the acute myeloid leukemias. Cancer Genet Cytogenet 1988; 30:1-15.

9. Drexler HG. Classification of acute myeloid leukemias: A comparison of FAB and immunophenotyping. Leukemia 1987; 1 (10):697-705.

10. Cheson BD, Cassileth PA, Head DR et al. Report of the National Cancer Institute-Sponsored Workshop on definitions of diagnosis in acute myeloid leukemia. J Clin Oncol 1990; 8(5):813-9.

11. Bene MC, Castoldi G, Knapp W et al. Proposals for the immunological classification of acute leukemias. European Group for the Immunological Characterization of Leukemias (EGIL). Leukemia 1995; 9:1783-6.

12. Harris NL, Jaffe ES, Diebold J et al. The World Health Organization classification of neoplastic diseases of the heamatopoietic and lymphoid tissues: report of the Clinical Advisory Committee Meeting, Airlie House, Virginia, november 1997. Histopathology 2000; 36:69-87.

13. Borowitz MJ, Guenther KL, Schults KE et al. Immunophenotyping of acute leukemia by flow cytometric analysis. Use of CD45 and right angle light scatter to get on leukemic blasts in three-color analysis. Am J Clin Pathol 1993; 100(5):534-40.

14. Stelzer GT, Shults Ke, Loken MR. CD45 gating for routine flow cytometric analysis of human bone marrow specimens. Ann NY Acad Sci 1993; 677:265-280.

15. Jennings CD, Foon KA. Recent advances in flow cytometry: Application to the diagnosis of hematologic malignancy. Blood 1997; 90 (8):2863-2892.

16. Bene MC, Bernier M, Castoldi G et al. Impact of immunophenotyping on management of acute leukemias. Haematologica 1999; 84(11):1024-34.

17. Fizzera G, Wu D, Inghirami G. The usefulness of immunophenotypic and genotypic studies in the diagnosis and classification of hematopoietic and lymphoid neoplasms. Am J Clin Pathol 1999; 111 (S1):13-39.

18. Pui CH, Behm FG, Crist WM. Clinical and biological relevance of immunologic marker studies in childhood acut lymphoblastic leukemia. Blood 1993; 82(2):343-62.

19. Loken MR, Wells DA. Normal antigen expression in hematopoiesis basis for interpreting leukemia phenotypes. In: Carleton CS, Nicholson JKA, eds. Immunophenotyping. New-York: Wiley-Liss Inc., 2000:133-160.

20. Davies BH, Foucar K, Szczarkowski W et al. US-Canadian Consensus recommendations on the immunophenotypic analysis of hematologic neoplasia by flow cytometry: Medical indicatons. Cytometry 1997; 30:249-263.

21. Sperling C, Steibt-Jung H, Gassmann W et al. Immunophenotyping of acute myeloid leukemia: Correlation with morphological characteristics and therapy response. Recent Results Cancer Res 1993; 131:381-92.

22. Hurwitz CA, Loken MR, Graham ML et al. Asynchronous antigen expression in B linegae acute lymphoblastic leukemia. Blood 1988; 72:299-307.

23. Terstrappen LW, Strafford M, Konemann S et al. Flow cytometric characterization of acute myeloid leukemia. Part II. Phenotypic heterogeneity at diagnosis. Leukemia 1991; 6:70-80.

24. Greaves M, Chan L, Furley A et al. Lineage promiscuity in hemopoietic differentation and leukemia. Blood 1986; 67:1-11.
25. Hurwitz CA, Mirro J Jr. Mixed-lineage leukemia and asynchronous antigen expression. Hematol Oncol Clin North Am 1990; 4:767-94.
26. Borowitz MJ, Bray R, Gascoyne R et al. US-Canadian Consensus recommendations on the immunophenotypic analysis of hematologic neoplasia by flow cytometry: Data analysis and interpretation. Cytometry 1997; 30:236-244.
27. Behm FG, Raimondi SC, Schell MJ et al. Lack of CD45 antigen on blast cells in childhood acute lymphoblastic leukemia is associated with chromosomal hyperploidity and other favorable prognostic features. Blood 1992; 79(4):1011-6.
28. Borowitz MJ, Shuster J, Carroll AJ et al. Prognostic significance of fluoresence intensity of surface marker expression in childhood B-precursor acute lymphoblastic leukemia. A Pediatric Oncology Group study. Blood 1997; 89:3960-6.
29. Janossy G, Coustan-Smith E, Campana D. The reliability of cytoplasmic CD3 and CD22 antigen expression in the immunodiagnosis of acute leukemia: a study of 500 cases. Leukemia; 1989; 3:170-81.
30. Legrand O, Perrot JY, Baudard M et al. The immunophenotype of 177 adults with acute myeloid leukaemia: proposal of a prognostic score. Blood 2000; 96:870-7.
31. Campana D, Behm FG. Immunophenotyping of leukemia. J Immunol Methods 2000; 243(1-2):59-75.
32. Pui CH, Frankel LS, Carroll AJ et al. Clinical characteristics and treatment outcome of childhood acute lymphoblastic leukemia with the t(4;11)(q21;q23): A collaborative study of 40 cases. Blood 1991; 77 (3):440-7.
33. Drexler HG, Thiel E, Ludwig W. Review of the incidence and clinical relevance of myeloid antigen positive acute lymphoblastic leukemia. Leukemia 1991; 5:637-45.
34. Reading CL, Estey EH, Huh YO et al. Expression of unusual immunophenotype combinations in acute myelogenous leukemia. Blood 1993; 81:3083-90.
35. Drexler HG, Thiel E, Ludwig WD. Acute myeloid leukemias expressing lymphoid associated antigens: diagnostic incidence and prognostic significance. Leukemia 1993; 7:489-98.
36. Venditti A, Del Poeta G, Buccisano F et al. Prognostic relevance of the expression of Tdt and CD7 in 335 cases of acute myeloid leukemia. Leukemia 1998; 12(7):1056-63.
37. Saxena A, Sheridan DP, Card RT et al. Biologic and clinical significance of CD7 expression in acute myeloid leukemia. Am J Hematol 1998; 58:278-84.
38. De Nully Brown P, Jurlander J, Pedersen-Bjergaard J et al. The prognostic significance of chromosomal analysis and immunophenotyping in 117 patients with de novo acute myeloid leukemia. Leuk Res 1997; 21(10):985-95.
39. Stewart CC, Behm FG, Carey JL et al. US-Canadian consensus recommendations on the immunophenotypic analysis of hematologic neoplasia by flow cytometry: selection of antibody combinations. Cytometry 1998; 30:231-5.
40. Rothe G, Schmitz G for the Working Group on Flow Cytometry and Image analysis and members of the editorial committee: Adorf D, Barlage S, Gramatzki M et al. Consensus protocol for the flow cytometric immunophenotyping of hematopoietic malignancies. Leukemia 1996; 10:877-95.
41. Ong ST, Larson RA. Current management of acute lymphoblastic leukemia in adults. Oncology 1995; 9(5):433-42.
42. Farahat N, Lens D, Morilla R et al. Differential TdT expression in acute leukemias by flow cytometry: a quantitative study. Leukemia 1995; 9:583-7.
43. Astsaturov IA, Matutes E, Morilla R et al. Differential expression of B29 (CD79b) and mb-1 (CD79a) proteins in acute lymphoblastic leukemia. Leukemia 1995; 10:769-73.
44. Borowitz MJ, Rubnitz J, Nash D et al. Surface antigen phenotype can predict TEL-AML1 rearrangements in childhood B-precursor ALL: A Pediatric Oncology Group Study. Leukemia 1998; 12:1764-70.
45. Boucheix C, David B, Sebban C et al. For the French Group on therapy for adult lymphoblastic leukemia: Immunophenotype of adult acute lymphoblastic leukemia, clinical parameters, and outcome: An analysis of a prospective trial including 562 tested patients (LALA87). Blood 1994; 84(5):1603-12.

46. Matutes E, Morilla R, Farahat N et al. Definition of acute biphenotypic leukemia. Haematologica 1997; 82:64-6.

47. Borowitz MJ, Hunger SP, Carroll AJ et al. Predictability of the t(1;19)(q23;p13) from surface antigen phenotype: Implications for screening cases of childhood acute lymphoblastic leukemia for molecular analysis. Blood 1993; 82:1086-91.

48. Pui CH, Hancock ML, Head DR et al. Clinical significance of CD34 expression in childhood acute lymphoblastic leukemia. Blood 1993; 82(3):889-94.

49. Koehler M, Behm FG, Shuster J et al. Transitional pre-B cell acute lymphoblastic leukemia of childhood is associated with favorable prognostic features and an excellent outcome: A Pediatric Oncology Group study. Leukemia 1993; 7:2064-8.

50. Navid F, Mosijczuk AD, Head DR et al. Acute lymphoblastic leukemia with the (8;14) (q24:q32) translocation of FAB L3 morphology associated with a B-precursor immunophenotype: The Pediatric Oncology Group Experience. Leukemia 1999; 13:135-41.

51. Uckun FM, Steinherz PG, Sather H et al. CD2 antigen expression on leukemic cells as a predictor of event-free survival after chemotherapy for T lineage acute lymphoblastic leukemia: A Children's Cancer Group Study. Blood 1996; 88:4288-95.

52. Shuster JJ, Falletta JM, Pullen DJ et al. Prognostic factors in childhood T-cell acute lymphoblastic leukemia: A Pediatric Oncology Group study. Blood 1990; 75(1):166-73.

53. Solary E, Casasnovas RO, Campos L et al. Surface markers in adult myeloblastic leukemia: correlation of CD19+, CD34+ and CD14+/DR- phenotypes with shorter survival. Groupe d' Etude Immunologique des Leucémies (GEIL). Leukemia 1992; 6:393-9.

54. Bennett JM, Catovsky D, Daniel MT et al. Proposal for the recognition of minimally differentiated acute myeloid leukemia (AML-M0). Br J Hematol 1991; 78:325-9.

55. Kotylo PK, Seo IS, Simth FO et al. Flow cytometric immunophenotypic characterization of pediatric and adult minimally differentiated acute myeloid leukemia (AML-M0). Am J Clin Pathol 2000; 113:193-200.

56. Venditti A, Del Poeta G, Buccisano F et al. Minimally differentiated acute myeloid leukemia (AML-M0): comparison of 25 cases with other French-American-British subtypes. Blood 1997; 89(2):621-9.

57. Kantarjian HM, Hirsch-Ginsberg C, Yee G et al. Mixed-lineage leukemia revisited: acute lymphocytic leukemia with myeloperoxidase-positive blasts by electron microscopy. Blood 1990; 76:808-13.

58. Cuneo A, Ferrant A, Michaux JL et al. Cytogenetic profile of minimally differentiated (FAB M0) acute myeloid leukemia: correlation with clinicobiologic findings. Blood 1995; 85 (12):3688-94.

59. Paloczi K. Application of monoclonal antibodies to the diagnosis and classification of acute leukemias. In: Paloczi K, ed. Immunophenotyping. 1st ed. Austin: RG Landes, 1994.

60. Lo Coco F, Avvisati G, Diverio D et al. Rearrangements of the RARα gene in acute promyelocytic leukemia: Correlations with morphology and immunophenotype. Br J Haematol 1991; 78:494-9.

61. Guglielmi C, Martelli MP, Diverio D et al. Immunophenotype of adult and childhood acute promyelocytic leukemia: corrleation with morphology, type of PML gene breakpoint and clinical outcome. A cooperative Italian study on 196 cases. Br J Haematol 1998; 102:1015-41.

62. Warrell RP, Maslak P, Eardley A. Treatment of acute promyelocytic leukemia with all-trans-retinoic acid: An update of the New York experience. Leukemia 1994; 8(6):929-33

63. Licht JD, Chomienne C, Goy A et al. Clinical and molecular characterization of rare syndrome of acute promyelocytic leukemia associated with translocation (11;17). Blood 1995; 85:1083-94.

64. Scott AA, Head DR, Kopeczky KJ et al. HLA-DR-CD33+, CD56+. CD16-, myeloid/natural killer cell acute leukemia: A previously recognised form of acute leukemia potentially misdiagnosed as French-American-British acute myeloid leukemia-M3. Blood 1994; 84(1):244-55.

65. Adriaansen HJ, te Boekhorst PAW, Hagemeijer AM et al. Acute myeloid leukemia with bone marrow eosinophilia (m4EO) and inv (16)(p13q22) exhibits a specific immunophenotype with CD2 expression. Blood 1993; 81 (11):3043-51.

66. Cuneo A, Van Orshoven A, Michaux JL et al. Morphologic, immunologic and cytogenetic studies in erythroleukemia: Evidence for multilineage involvement and identification of two distinct cytogenetic-clinicopathologic types. Br J Haematol 1990; 75 (3):346-54.

67. Betz SA, Foucar K, Head DR et al. False-positive flow cytometric platelet glycoprotein IIb/IIIa expression in myeloid leukemias secondary to platelet adherence to blasts. Blood 1992; 79 (9):2399-403.

68. Norton JD, Campana D, Hoffbrand AH et al. Rearrangement of immunoglobulin and T cell antigen receptor genes in acute myeloid leukemia with lymphoid-associated markers. Leukemia 1988; 1:757-61.

69. Killick S, matutes E, Powles RL et al. Outcome of biphenotypic acute leukemia. Haematologica 1999; 84:699-706.

70. Buccheri V, Matutes E, Dyer MJ et al. Lineage commitment in biphenotypic acute leukemia. Leukemia 1993; 7 (6):919-27.

71. Campana D, Yokota S, Coustan-Smith E et al. The detection of residual acute lymphoblastic leukemia cells with immunologic methods and polymerase chain reaction: a comparative study. Leukemia 1990; 4:609-14.

72. Campana D, Freitas RO, Coustan-Smith E. Detection of residual leukemia with immunologic methods: Technical developments and clinical implications. Leukemia Lymphoma 1994; 13(S1):31-34.

73. Campana D, Pui CH. Detection of minimal residual disease in acute leukemia: Methodologic advances and clinical implications. Blood 1995; 85:1416-34.

74. Campana D, Coustan-Smith E. Detection of minimal residual disease in acute leukemia by flow cytometry. Cytometry 1999; 38(4):139-52.

75. Adriaansen HJ, Jacobs BC, Kappers-Klunne MC et al. Detection of residual disease in AML patients by use of double immunologic marker analysis for terminal deoxynucleotidyl transferase and myeloid markers. Leukemia 1993; 7:472-81.

76. Ciudad J, San Miguel JF, Lopez-Berges MC et al. Immunophenotypic detection of minimal residual disese in acute lymphoblastic leukemia. In: Büchner T, Hiddemann W, Wormann B et al. eds. Acute leukemias Vol VI. Prognostic factors and treatment strategies. 1st ed. Berlin: Springer-Verlag, 1997:321-7.

77. Coustan-Smith E, Behm F, Sanchez J et al. Immunological detection of minimal residual disease in children with acute lymphoblastic leukemia. Lancet 1998; 351:550-4.

78. Macedo A, Orfao A, Gonzales M et al. Immunological detection of blast cell subpopulations in acute myeloblastic leukemia at diagnosis: implications for minimal residual deisease studies. Leukemia 1995; 9:993-8.

79. Liu Yin JA. Detection of minimal residual disease in acute myeloid leukemia. methodologies, clinical and biological significance. Br J Haematol 1999; 106:578-90.

80. Te Boekhorst PA, de Leeuw K, Schoester M et al. predominance of functional multidrug resistance (MDR-1) phenotype in CD34+ acute myeloid leukemia cells. Blood 1993; 82(10):3157-62.

81. Willmann CL. The prognostic significance of the expression and function of multidrug resistance transporter proteins in acute myeloid leukemia: Studies of the Southwest Oncology Group Leukemia Reseach Program. Semin Hematol 1997; 4(S1):25-33.

82. Karaszi E, Jakab K, Homolya L et al. Calcein assay for multidrug resistance reliably predicts therapy response and survival rate in acute myeloid leukemia. Br J Haematol 2001; 112(2):308-14.

83. Pall G, Spitaler M Hofmann J. Multidrug resistance in acute leukemia. a comparison of different diagnostic methods. Leukemia 1997; 11:1067-72.

CHAPTER 3

Immunophenotypic Markers in the Diagnosis of Non-Hodgkin's Lymphomas

András Matolcsy

N on-Hodgkin's lymphomas (NHL) represent a large, clinically, morphologically and biologically diverse spectrum of malignant neoplasms. The classification and diagnosis of NHLs are a traditionally difficult problem for both pathologists and hematologists. One circumstance accounts for the difficulties is that neoplastic lymphoid cells often share morphologic, imunophenotypic and molecular features with those of normal lymphoid cells. The second circumstance is the lack of complete knowledge and understanding the maturation and differentiation of the normal hematopoietic and immune system. The new scientific results that has appeared during the past decades on the hematopoietic and immune system have generated time to time scientific basis for the discovery of new lymphoma entities and created new lymphoma classifications. In 1994 the International Lymphoma Study Group published a consensus proposal for the classification of lymphoid neoplasms named Revised European American Lymphoma (REAL) classification.[1] This is based on the principle that the classification is a list of "real" entities defined by the combination of morphology, immunophenotype, genetic and clinical features. Since 1995, members of the European and American Hematopathology Societies have been collaborating on a new WHO classification of lymphomas. This classification is an updated version of REAL representing the first international consensus on the classification of lymphomas.[2]

The immunophenotypic profiles of each lymphoma entities have been integrated in the diagnosis and classification of lymphomas. The widespread application of immunophenotypic analysis in lymphomas is due to the standardization of cytofluorometric and immunohistochemical techniques and the development of well-characterized monoclonal antibody reagents. These factors have resulted in a voluminous data of the literature concerning the reactivity of different monoclonal antibodies with normal and malignant lymphoid cells.[3] In this chapter the immunophenotypic profiles of the major lymphoma entities will be discussed which have been defined by the REAL/WHO classification of lymphomas.

Lineage of Non-Hodgkin's Lymphomas

In the REAL/WHO classification the major lymphoma categories have been separated according to the putative normal counterpart cells. Based on these principles NHLs are separated into B- and T/NK-cell categories. The determination of lineage commitment of lymphomas is basically on the immunophenotypic analysis of the tumor cells. There is no any gold standard for the determination of cellular lineage of lymphomas, but combination of different monoclonal antibodies is helpful to find out the cellular origin of lymphomas.

The most frequently used reagents to detect B-cell lineage with the greatest specificity and sensitivity are the CD19, CD20 and CD22 clusters.[3] In formoll-paraffin embedded tissues, L26 monoclonal antibody (MoAb) recognizing an intracellular epitope of CD20 and MoAb JCB117 detecting CD79a antigen expressing by B-cells in association with surface immunglobulin are the most helpful.[4,5] The advantages of CD79a over CD20 is that CD79a is expressed by normal and neoplastic B-cells at all stages of differentiation from progenitor B-cell to plasma cell. A diagnosis of B-cell lymphoma requires evidences of monoclonality. Monoclonality of a B-cell proliferation can be verified by detection of κ or λ light chain restriction using immunohistochemical analysis or by the detection of clonal immunoglobulin gene rearrangement using molecular studies.[6]

The immunophenotypic identification of T-cell lineage is based on the detection of one or more pan-T-cell markers (CD2, CD3, CD5 and CD7), in the absence of B-cell associated antigens.[3,7] However, the immunophenotypic criteria for T-cell neoplasia are based on other antigen profiles including T-cell subset antigen restriction, anomalous T-cell subset antigen expression and a precursor T-cell immunophenotype.[8] In a marked T-cell proliferation with restriction of CD4+ or CD8+ cells raises the possibility of T-cell neoplasm but these are not sufficient to warrant a definitive diagnosis of lymphoma. Anomalous T-cell subset antigen expression is a more useful criterion in the immunodiagnostics of T-cell neoplasia. The presence of CD4+/CD8+ or CD4-/CD8- T-cell population is distinctly abnormal and suggestive for T-cell neoplasia. Finally, the detection of TdT and CD1 on peripheral T-cells which antigens usually expressed in various stages of intrathymic T-cell differentiation is abnormal that also suggestive for T-cell neoplasia. Although these immunphenotypic criteria are useful in the diagnosis of T-cell neoplasia, no immunophenotypic marker of T-cell clonality exist, unlike immunoglobulin light chain restriction in clonal B-cell proliferations.

The identification of NK cell lineage is based on the detection of CD56, CD57 and CD16 epitopes. These antigens are frequently co-expressed with other T-cell antigens including CD2, CD3, CD7 and CD8 because T- and NK-cells are closely related lineages originating in a common T/NK-cell progenitor.[9] The presence of intracellular cytotoxic molecules like granzyme-B, perforin and/or TIA1 also help to identify lymphoproliferations with NK- or cytotoxic T-cell origin.[10]

B-Cell Non-Hodgkin's Lymphomas

Precursor B-Lymphoblastic Leukemia/Lymphoma

Precursor B-cell lymphomas presenting as solid tumors (B-lymphoblastic lymphoma) and those of presenting with bone marrow and blood involvement (B-lymphoblastic leukemia) are biologically the same diseases with different clinical presentations. Since most precursor lymphoid neoplasms present as leukemia, the term acute lymphoblastic leukemia (ALL) has been retained for the leukemic phase of precursor B-cell neoplasms.[2] Precursor B-lymphoblastic leukemia/lymphoma (B-LBL) is composed of lymphoblasts that are slightly larger than small lymphocytes with round or convoluted nuclei, fine chromatin and basophilic cytoplasm. In solid tissue involvement the number of mitoses are high and a "starry sky" pattern frequently present.[1]

Tumor cells of B-LBL characteristically express TdT, CD19, CD79a and HLA-DR but no surface immunoglobulin. In variable number of cases, CD10, CD20, CD22, CD34 and cytoplasmic IgM also can be expressed.[1,3] Cases with CD10 expression have better prognosis than cases without CD10. Expression of CD20 and cytoplasmic IgM increases with maturity of the tumor cells. Among patient with B-LBL several genetic subgroups can be identified that highly correlate with prognosis[11] (Table 1). Certain cytogenetic alterations may also be associated with distinct phenotype of the tumor cells. Cases with translocation of chromosome 11q23 are frequently CD10 negative. Both the t(9;22) and t(11q23) genetic abnormalities may associate with expression of myeloid-associated antigens (CD13, CD33) on B-cells.[12]

Table 1.   Cytogenetic subgroups of precursor B-lymphoblastic leukemia/lymphoma and their clinical prognosis

Cytogenetic Abnormality	Molecular Alteration	Prognosis
t(9;22)(a34;q11)	BCR/ABL	Poor
t(v;11q23)	MLL rearranged	Very poor
t(1;19)(q23;q13)	E2A/PBX1	Poor
t(12;21)(p12;q22)	ETV/CBF- alpha	Excellent

Chronic Lymphocytic Leukemia/Small Lymphocytic Lymphoma

Chronic lymphocytic lymphoma/small lymphocytic lymphoma (CLL/SLL) is a clonal lymphoproliferative disorder of B-cell origin that manifest as a progressive accumulation of mature, non-activated B-cells in the blood, bone marrow and lymphoid tissues. Most cases associated with leukemic blood picture, and only rare cases have isolated lymph node involvement called SLL. Histologically, CLL/SLL shows a monotonous infiltration of small cells in the lymphoid organs, but clusters of larger cells of the lymph nodes, called pseudofollicules or proliferation centers are a common feature.[1] CLL and SLL are biologically related very closely to each other, therefore the REAL/WHO classifications has eliminated SLL as distinct category and has submitted under the category of CLL.[2]

CLL/SLL derives from the neoplastic transformation of normal CD5+ (B1) B-cells. Tumor cells of CLL/SLL express low-density monotypic surface immunoglobulin, usually IgM or IgM and IgD but CD79b molecule is lacking.[13] B-cell associated antigens CD19, CD20, CD23, CD43, and CD79a are co-expressed with CD5 in almost all cases.[3,14] CD10 and CD22 are negative, whereas CD11c and CD25 are often expressed, but tend to be weak.[3,15] CLL/SLL cells express bcl-2 and lack bcl-6 gene expression. The CD20 and CD23 antibodies have stronger reaction in the pseudofollicules than in the small cells. In the pseudofollicules delicate follicular dendritic cell network is detectable by CD21. Ki-67 positive cells are more frequent in the pseudofollicules but the positivity is usually below 20%. By using this broad antigenic fingerprint of the leukemic cells, CLL/SLL can usually be distinguished from other mature B-cell neoplasms such as mantle cell lymphoma, follicle center cell lymphoma, splenic lymphoma with villous lymphocytes and hairy cell leukemia.[3,14,15]

Several attempts have been made to identify possible immunologic markers associated with more aggressive clinical behavior or poor prognosis. Among them adhesion molecule expression (CD44, CD58) may have clinical significance.[16] More recent studies suggest that patients with chromosomal abnormalities like trisomy 12 and abnormalities of 14q, 13q and 22q tend to have a poorer prognosis.[17]

In approximately 3-10% of CLL/SLLs, diffuse large B-cell lymphoma (Richter's syndrome) or Hodgkin's lymphoma may develop. In these transformed cases the neoplastic cells may share immunophenotype of the original CLL/SLL cells or alternatively they can be different. In those cases where the immunophenotype of the CLL/SLL and the transformed tumors are different the two lymphomas may have different clonal origin.[18]

Prolymphocytic Leukemia

Prolymphocytic leukemia (PLL) is typically a more aggressive disease than CLL/SLL. It may develop through transformation of CLL/SLL or alternatively as de novo disease.[1,19] Morphologically, PLL cells are larger than cells of CLL/SLL they have wider cytoplasm, clumped chromatin and prominent single and central nucleoli. If in patients with CLL/SLL the number of PLL cells exceeds 30%, it is called prolymphocytic transformation of CLL/SLL. In prolymphocytic transformation the tumor cells may retain the immunophenotypic character-istics of the earlier CLL/SLL or demonstrate brighter surface immunoglobulin, as well as vari-able CD5 and FMC7 positivity.[1,19]

Lymphoplasmacytic Lymphoma/Immunocytoma

Lymphoplasmocytoid lymphoma/immunocytoma (LPL/IC) is an indolent B-cell neo-plasm, which composed of small B-lymphocytes with a characteristic plasma cell differentia-tion. The disease usually involves the bone marrow, lymph nodes and spleen. Patients may develop serum monoclonal protein with hyperviscosity (Waldenstrom's macroglobulinaemia). The plasmocytic elements frequently show intranuclear immunoglobulin inclusions (Dutcher-bodies). The pattern of lymph node involvement is usually diffuse, the bone marrow involvement either diffuse or nodular. The peripheral blood involvement is usually less promi-nent than in CLL, and the cells often have a plasmacytoid appearance.[1]

The LPL/IC cells express surface and sometimes cytoplasmic IgM and lack IgD. Tumor cells strongly express the B-cell associated antigens CD19, CD20, CD22 and CD79a but not the CD10 or CD23. The CD5 expression is absent.[1,15,21] Approximately, 40-50% of cases carry the t(9;14)(p13;q32) chromosomal translocation, that juxtaposes the PAX-5 gene to the IgH gene. The translocation results overexpression of the PAX-5 protein.[22]

Mantle Cell Lymphoma

Mantle cell lymphoma (MCL) is a B-cell non-Hodgkin's disorder that represents 2-8% of lymphomas.[23] Most patients with MCL are older and present with generalized lymphadenopa-thy and frequent involvement of spleen and bone marrow. Leukemic presentation also may occur. Extranodal disease occurs in the gastrointestinal tract as multiple lymphomatous polyposis.[24]

Histologically, MCL consists of atypical small lymphoid cells and have either a nodular or diffuse pattern of growth or the combination of the two patterns. In nodular MCL, nodules may consist of follicles with reactive germinal centers surrounded by broad mantles of small neoplastic cells, the so-called mantle zone pattern. In diffuse growth pattern, invasion and obliteration of the reactive germinal centers and interfollicular areas by neoplastic cell result in an effacement of the nodal architecture. Cytologically, neoplastic cells of MCL may show dif-ferent forms. In classical variant of MCL, the tumor is composed of small lymphoid cells with irregular and indented nuclei, moderately coarse chromatin and inconspicuous nucleoli. The high-grade variant of MCL also called blastic variant, composed of more pleiomorphic tumor cells. In these cases, the tumor cells are medium to large in size, with fine chromatin and multiple small nucleoli. The blastic variant of MCL may occur de novo or as transformed form of classical MCL.[25]

The immunophenotype of MCL is consistent with that of the normal mantle zone cells with the exception of CD23. Tumor cells express surface IgM and frequently IgD. The expres-sion of λ light chain is more frequent than κ. The tumor cells are co-expressing a variety of pan-B cell antigens including CD19, CD20, CD22, CD43 with the CD5 antigen. The tumor cells are negative for CD23, CD10 and CD11c. Antibodies to follicular dentritic cells (CD21) reveal aggregates of these cells in nodular type of MCL. Blastic variant of MCL may show growth fraction with Ki-67 beyond 30% and may express CD10 antigen.[1,25] The overexpression of p53 does correlate with disease progression.

The most helpful and sensitive immunohistochemical staining in the diagnosis of MCL the detection of the overexpression of the PRAD-1/cyclin D1 protein by the tumor cells, which is highly characteristic for this type of lymphoma. The over expression of the cyclin D1 is related to the specific cytogenetic alteration t(11;14)(q13;q32), which result in rearrangement and deregulation of the bcl-1/PRAD-1 proto-oncogene.[26]

Follicular Lymphoma

Follicular lymphoma (FL) is a tumor of germinal center B-cells composed of centrocytes (small cleaved cells) and centroblasts (large noncleaved cells). The proportion of centrocytes and centroblasts vary among cases. The clinical aggressiveness of the tumor increases with increasing numbers of centroblasts. In addition to cellular composition, the proportion of follicular and diffuse areas is varying from case to case, and these are also associated with prognosis.[1]

FL cells express the CD19, CD20, CD22 and CD79a pan-B-cell antigens, about 60% of the cases also express CD10 and they are negative for CD5, CD43 and CD11c antibodies. In more than 50% of cases, tumor cells express sIgM, about 40% express sIgG and rare cases express IgA. Tightly organized meshwork of follicular dendritic reticulum cells is detectable with CD21 antibodies.[15] In most cases, FL cells express the bcl-2 protein, however bcl-2 positivity of the tumor cells can be lost in the clinical course and high-grade transformation of the tumor.[27] The bcl-6 nuclear protein also expressed in a significant fraction of the tumor cells.[28]

Approximately 85% of FLs are associated with the t(14;18) translocation that places the bcl-2 oncogene into juxtaposition with the joining (J$_H$) gene segment cluster of the immunoglobulin (Ig) heavy chain (H) gene locus. The t(14;18) translocation upregulates the expression of the bcl-2 gene-product which induces prolonged cell survival by blocking programmed cell death (apoptosis).[29]

Nodal Marginal Zone B-Cell Lymphoma

Nodal marginal zone B-cell lymphoma (MZL) is characterized by heterogeneous proliferation of atypical neoplastic B-cells that includes monocytoid B-cells, small to medium-sized cells with irregular nuclei (centrocyte-like cells) and variable number of plasma cell. The tumor cells infiltrate predominantly the parafollicular and perisinusoidal areas but diffuse obliteration of the nodal architecture may occur as well. In some cases leukemic phase of MZL has been reported but these complication is extremely rare.

Nodal MZL is a mature B-cell neoplasm that expresses B-cell associated antigens CD19, CD20, CD22 and CD79a, but other B-cell antigens like CD10, CD23, CD21 and CD35 are usually absent. MZL is usually not associated with CD5 reactivity. Tumor cell express frequently surface IgM but generally lack IgD. In rare cases, IgG and IgA expression has been documented. Monotypic cytoplasmic immunoglobulin may occur in the perinuclear space of tumor cells. Bcl-2 immunostaining may be observed in 50-80% of cases and bcl-6 expression is absent in low-grade MZL. The DBA44 monoclonal antibody that strongly reactive in hairy cell leukemia could be positive in MZL, occasionally.[30]

Cytogenetic and molecular studies show many similarities in nodal and extranodal MZLs. Trisomy 3 has been found as most frequent genetic alterations in nodal MZLs, but chromosomal translocation t(11;18)(q21;q21) is no characteristic for nodal MZLs.[31]

Extranodal Marginal Zone B-Cell Lymphoma of MALT-Type

When marginal zone B-cell lymphomas occurs at extranodal sites usually on glandular epithelial tissues (gastrointestinal tract, lung, salivary glands, thyroid glands etc.) it is commonly referred to as MALT lymphoma. These neoplasms derive from B-cells associated with epithelial tissues. MALT lymphoma is characterized by a polymorphous infiltrate of small lymphocytes, marginal zone (centrocyte-like) B cells, monocytoid B cells and plasma cells. In epithelial tissues, tumor cells typically infiltrate the epithelium, forming so-called lymphoepithelial lesions. Tumor cells usually surround reactive follicules occupying the marginal zone or the interfollicular region. The invasion of reactive follicules by the neoplastic cells (follicular colonization) is a frequent histological feature of MALT lymphoma. High-grade transformation of MALT lymphomas is a relatively frequent consequence of the disease.[32]

The tumor cells express most frequently surface IgM, followed by IgG and IgA but lack IgD. In 40-60% of cases monotypic cytoplasmic immunglobulin is detectable indicating plasmacytoid differentiation. They express CD19, CD20, CD22 and CD79a B-cell associated antigens and are CD10, CD23 and CD5 negative.[1,15] The number of Ki-67 positive cells rises significantly upon transformation to a diffuse large B-cell lymphoma.

Trisomy 3 is found in approximately 60% of cases, chromosome translocation t(11;18)(q21;q21) in found in around 25-40%.[33,34] Chromosomal translocation t(1;14)(p22;q32) which involve an apoptosis-promoting gene (bcl-10) is rare and appears to be associated with more disseminated or aggressive tumor.[35]

Splenic Marginal Zone B-Cell Lymphoma

Splenic marginal zone B-cell lymphoma (SMZL) is an indolent lymphoma arises from the marginal zone B-cells in the spleen. Tumor cells of SMZL use to involve red pulp and both mantle and marginal zone of the splenic white pulp and leave intact central residual germinal centers. Patients typically have bone marrow involvement and leukemic blood picture without lymph node involvement. The disease corresponds to a unique chronic leukemia known as "splenic lymphoma with villous lymphocytes".[1,36]

The immunophenotype of SMZL is similar to that of extranodal and nodal marginal zone B-cell lymphoma. Tumor cell express both surface IgM and IgD which is characteristic for the splenic-type of MZL in contrast to the MALT-type where surface IgD is usually absent. Tumor cells express the B-cell associated antigens CD19, CD20, CD22 and CD49a, and lack CD5, CD10, CD11c, CD23 and CD25.[1,36]

Hairy Cell Leukemia

Hairy cell leukemia (HCL) is a mature B-cell neoplasm characterized by pancytopenia, myelofibrosis and splenomegaly. The neoplastic cells are small with oval or bean-shaped nuclei, and abundant pale cytoplasm with "hairy" membrane projections on their surface. They are preferentially accumulated in the bone marrow and in the splenic red pulp. In the bone marrow infiltrated by the neoplastic cells, the reticulin fiber is increased, often resulting in a "dry tap". The diagnosis most frequently is made through bone marrow biopsy.[1]

The immunophenotype of tumor cells shows unique features of B-cells. They are express B-cell associated antigens CD19, CD20, CD22 and CD79a and they are negative with CD5, CD10, CD23 and CD43. HCL cells are expressing several other antigens detecting by MoAbs including α chain of the leukocyte adhesion molecule p150 (CD11c), low affinity p55 receptor for interleukin-2 (CD25) and an integrin subunit β_7 (CD103). Unfortunately, these epitopes are not detectable in formoll-paraffin embedded tissues. MoAb DBA.44 stains HCL cells even in decalcified paraffin-embedded tissues.[30,37] Tartarat-resistant acid phosphatase is present in most cases however, it is not specific for the diagnosis because this enzyme can be present in other type of disorders than HCL.

Diffuse Large B-Cell Lymphoma

Diffuse large B-cell lymphoma (DLBL) is a neoplasm of transformed large B-cells with diffuse growth pattern and a high proliferation fraction (> 40%). DLBLs are a heterogeneous group of neoplasms. They are composed of large cells resemble centroblasts or immunoblasts most often with a mixture of the two. Several morphologic variants (centroblastic, immunoblastic, anaplastic, multilobated, T-cell or histiocyte rich, plasmoblastic, lymphomatoid granulomatosis) has been recognized but their clinical significance is debated.[1,2]

Generally, DLBLs express one or more B-cell associated antigens (CD19, CD20, CD22, CD79a) and CD45. Tumor cells may express monotypic surface or cytoplasmic immunoglobulins. Bcl-2 protein expression occurs in a 30-40% of the cases and bcl-6 expression present approximately 60% of the cases. A few cases of DLBL may show unique immunophenotype that suggestive for a distinct pathogenesis of these DLBLs. In rare cases, B-cell associated antigens are co-expressed with the CD5 epitope without history of CLL/SLL or Richter's syndrome. These cases are separated under the name of "de novo CD5 positive DLBL".[38] Other rare cases co-express the B-cell associated antigens with the CD30 molecule. These cases are believed to be an anaplastic large cell lymphoma of B-cell origin.[39] Furthermore, cases in where the relative numbers of reactive T-cells and/or histiocytes are high comparing to the DLBL cells are separated under the name of T-cell/histiocyte rich B-cell lymphoma.[40]

In the group of DLBLs three distinct clinicopathological entities have been distinguished based on their clinical manifestation and separated as subtypes of DLBL.

1. Primary mediastinal B-cell lymphoma (PMBCL): PMBL is an aggressive tumor of the anterior mediastinum that usually involves the thymus. Tumor cells are usually large cells with clear cytoplasm. Many cases have fine, compartmentalizing sclerosis. The tumor cells share the immunophenotype of DLBLs and occasionally they express CD30 as well.[41]

2. Intravascular large B-cell lymphoma (IVLBCL): IVLBCL is a rare, aggressive variant of DLBL in which neoplastic cells are disseminated in the vascular lumina. Tumor cell express one or more B-cell associated antigens and occasionally these antigens are co-expressed with CD10 or CD5.

3. Primary effusional lymphoma (PEL): PEL is an aggressive NHL that grows in the pleural, pericardial, and abdominal cavities as a lymphomatous effusion, usually in the absence of a detectable tumor mass. PEL occurs predominantly but not exclusively among human immunodeficiency virus (HIV)-infected individuals. These neoplasms exhibit cytomorphologic features that bridge large cell immunoblastic and anaplastic large cell lymphoma, usually lack surface immunoglobulin and B-cell-associated antigens but express CD45 and antigens associated with late stages of B cell differentiation such as CD30, CD38, CD71, CD138, and epithelial membrane antigen (EMA). Genotypic analysis of the tumor cells has revealed clonal Ig gene rearrangement, Epstein-Barr virus (EBV) and Kaposi's sarcoma-associated herpesvirus (KHSV)/human herpesvirus-8 (HHV-8) infection in most cases.[42]

Plasma Cell Myeloma/Plasmacytoma

Plasma cell myeloma/plasmocytoma is a group of related disorders associated with a malignant plasma cell proliferation involving the bone marrow (plasma cell myeloma) or extraosseal soft tissues (plasmacytoma). Plasma cell myeloma (plasmacytoma) is characterized by monoclonal gammopathy, osteolytic lesion and frequent renal failure.

Plasma cell myeloma/plasmacytoma express monotypic cytoplasmic immunglobulin and plasma cell-associated antigen CD38. Tumor cells lack B-cell associated antigens CD19, CD20, CD22 and surface immunglobulins but CD79a can be expressed in some cases. Tumor cell also express the collagen-1 binding proteoglycan, syndecan-1 (CD138). Occasionally, tumor cells may express immature B-cell antigens like CD10 and terminal deoxynucleotidyl transferase (TdT) that usually associated with poor clinical prognosis.[1]

Chromosomal translocation t(11;14) is found in 20-30% of cases but the breakpoints differ than that of MCL. In cases carry the translocation cyclin D1 is overexpressed.[43]

Burkitt's Lymphoma

Burkitt's lymphomas (BL) may occur in three different forms including African or endemic type, non-African or sporadic type and immunodeficiency associated type. In African type, the jaw and other facial bones are involved. In non-African type the majority of the cases ileum, cecum, mesentery and ovaries are involved. Rare cases present as acute leukemia as well. BL is an aggressive B-cell lymphoma consisting of medium-sized round cells with a loose chromatin pattern and relatively wide basophilic cytoplasm. The number of mitoses and apoptotic cells are high. The tumor has an extremely high rate of proliferation. Scattered macrophages with phagocytosed apoptotic cells give rise to the "starry sky" pattern with tumor cells of BL.

Tumor cells express B-cell associated antigens such as CD19, CD20, CD22, CD79a, as well as other B-cell antigens that frequently expressed in germinal center B-cells (CD10, CD38, CD40). In almost all cases, surface immunoglobulins, most frequently IgM and rarely IgG and IgA are present. A subset of BL is associated with Epstein-Barr virus (EBV) infection. In these cases CD21 is expressed that bind both complement (C3b) and EBV.

Most cases of BL carry the translocation of the c-myc oncogene. C-myc oncogene from the chromosome 8 juxtaposes to the immunglobulin heavy chain on chromosome 14 [t(8;14)] or to the light chain loci on chromosomes 2 [t(2;8)] or 22 [(8;22)].[44] EBV genomes can be demonstrated in tumor cells in most African type and less frequently in non-African and immunodeficiency associated types.

T-Cell Non-Hodgkin's Lymphomas

Precursor T-Lymphoblastic Lymphoma/Leukemia

T-lymphoblastic lymphoma/leukemia (T-LBL) is a neoplasm of precursor T-cells that may show two different clinical manifestations. It may occur as solid tumor mass usually involving the thymus or lymph nodes or as acute leukemia. Solid form of T-LBL may progress to acute leukemia in the course of the disease. Tumor cells are morphologically identical to those of B-LBL and immunophenotypic analysis is necessary to distinguish the two diseases.

In most cases, tumor cells are CD3 and CD7 positive and the expression of other T-cell associated antigens like CD2 and CD5 are variable. The tumor cells frequently show TdT and CD1a expression that consistent with the phenotype of early thymocytes. CD4 and CD8 can be double positive or double negative.[45] Occasionally, cases may express NK-cell antigens (CD16 and/or CD57) that frequently associated with more aggressive clinical course.[46] The T-cell genotype can be identified of T-LBL by the detection of clonal T-cell receptor gene rearrangement, however these cases may show clonal Ig gene rearrangement as well.[47]

T-Cell Prolymphocytic Leukemia

T-cell prolymphocytic leukemia (T-PL) is a rare aggressive leukemia characterized by proliferation of small- to medium-sized lymphocytes with prominent nucleoli and moderately abundant nongranular cytoplasm often with blebs. In some of the cases, tumor cells have cerebriform nuclei and inconspicuous nucleoli. Lymph node, splenic red pulp and hepatic sinusoids can be infiltrated. The lymph node and bone marrow involvements are diffuse.[1,7]

Tumor cells usually express T-cell associated antigens (CD2, CD3, CD5, and CD7) and the TCR $\alpha\beta$ chains. About 65% of the cases express CD4, 20% express both CD4 and CD8 and rare cases express only CD8.[7,48] Abnormalities of chromosome 14 with breakpoints at q11 and q32 are present in more than 90% of the cases.[1,3]

T-Cell Granular Lymphocytic Leukemia

T-cell granular lymphocytic leukemia (T-GLL) is a relatively indolent disease that generally infiltrates bone marrow and associated with leukemic dissemination. Lymph nodes are rarely involved but splenic red pulp and hepatic sinuses may be infiltrated. T-GLL cells are usually small and characteristically show a wide rim of pale blue cytoplasm that contains azurophilic granules.[1]

Based on the antigen expression profile of the tumor cells, T-suppressor and NK-cell types could be distinguish.[49] In T-suppressor cell type the tumor cells express CD2, CD3, CD8 and TCR $\alpha\beta$ but lack CD4. In come cases CD16 and/or CD57 may be positive. In NK-cell type, tumor cell also express T-cell associated antigens like CD2 but CD3, CD4, CD8 and TCR$\alpha\beta$

expression are absent. In these cases, the NK-cell associated antigens (CD16, CD56 and CD57) are expressd by the tumor cells. T-cell types carry clonal TCR gene rearrangement but NK-cell type has it in germ line.[50]

Aggressive NK-Cell Leukemia

Aggressive NK-cell leukemia is a neoplasm of immature NK-cells with bone marrow infiltration, leukemic blood picture, hepatosplenomegaly, lymphadenopathy and aggressive clinical course. The disease is extremely rare and mostly occurs in Asian. In some cases, nasal T/NK-cell lymphoma progress to aggressive NK-cell leukemia. Tumor cells are medium-sized with irregular hyperchromatic nucleoli and abundant cytoplasm that contain azurophilic granules. The number of mitoses is high.

Tumor cells mostly express NK-cell specific antigens like CD56, TIA-1 and granzyme B and some T-cell specific antigens (CD2, CD3ε). The tumor is always associated with EBV infection.[51,52]

Adult T-Cell Leukemia/Lymphoma

Adult T-cell leukemia/lymphoma (ATL/L) is a peripheral T-cell neoplasm that associated with infection of human T-cell lymphotrophic virus type I (HTLV1). Most cases occur in Japan and in the Caribbean, sporadic cases may occur elsewhere. Several clinical variants have been described such as acute, lymphomatous, chronic and smoldering depending on the clinical features. ATL/L encompass a morphologic spectrum of disorders. Tumor cells are usually varied from medium-sized to large with nuclear pleomorphism. Nuclei are commonly hyperlobated. Cytoplasm can be amorphophilic, basophilic or pale.

Tumor cells express T-cell associated antigens like CD2, CD3, CD5, but usually lack CD7. Most cases are CD4+ and CD8-. Rare cases are CD4- and CD8+ or CD4+ and CD8+. Tumor cells frequently express interleukin-2 receptor γ chain (CD25). The TCR gene is clonally rearranged and HTLV-1 genomes are integrated in all cases.[7,53,54]

Extranodal NK/T-Cell Lymphoma, Nasal Type

Extranodal NK/T-cell lymphoma, nasal type occurs in extranodal sites, predominantly in nose, nasopharinx and paranasal sinuses. Because the most frequent manifestation of the tumor is in the midfacial structures it was called "lethal midline granuloma". The majority of the tumor characterized by prominent angiocentric and angiodestructive growth pattern and marked necrosis. The cytology of the tumor cells is highly variable ranging from small- to medium-sized and occasionally large cells. Tumor cells are frequently admixed with inflammatory cells.

Extranodal NK/T-cell lymphoma, nasal type appears to be derived from NK cells, therefore the immunophenotype of tumor cells are similar to those of NK-cells. They express CD56 and TIA-1 but lack CD3, CD4 or CD8. The tumor cells are infected by EBV.[55]

Enteropathy-Type T-Cell Lymphoma

Enteropathy-type T-cell lymphoma (EATCL) is thought to be a neoplastic disease of the intraepithelial T-lymphocytes of the small intestine that usually associated with history of gluten-sensitive enteropathy. Most patient presents acute abdominal episodes due to small intestine perforation or obstruction. Tumor cell show a broad spectrum of morphological spectrum ranging from small to medium-sized and large cells. Anaplastic cells may be present.

EATCL express T-cell associated and/or NK-cell associated antigens. Most cases are CD43, CD3, CD7 and CD45RO positive, and some cases may also express CD8 and cytotoxic T-cell-associated proteins (granzyme B, TIA-1, perforin). Immunhistologically, characteristic finding is the expression of the mucosal homing receptor CD103 antigen. The TCR gene shows clonal rearrangement.[56]

Hepatosplenic Gamma-Delta T-Cell Lymphoma

Hepatosplenic gamma-delta T-cell lymphoma is a rare disease that characterized by the expression of the γ/δ complex. Most patients with hepatosplenic T-cell lymphomas are young who presents hepatosplenomegaly without lymphadenopathy. Peripheral blood and bone marrow may be involved in a small proportion of the patients. In the spleen, the infiltrate involves the red pulp and the sinuses. Tumor cells are medium-sized with round or oval nuclei, condensed chromatin and abundant pale to eosinophilic cytoplasm.

Neoplastic cells express T-cell associated antigens (CD2, CD3, CD5 and CD7) and TCR γ/δ molecule, but lack neither CD4 nor CD8. Tumor cells also express cytotoxic molecules (Perforin, Granzyme B, TIA-1) as well NK-cell associated molecules (CD16, CD56). The TCR γ and δ genes are clonally rearranged.[57]

Subcutaneous Panniculitis-Like T-Cell Lymphoma

Subcutaneous panniculitis-like T-cell lymphoma is characterized by multiple subcutaneous nodules generally involving the extremities and occasionally the trunk. The disease restricted to subcutaneous tissue and atypical tumor cells infiltrate between fat cells of subcutaneous tissue with no dermal involvement. Tumor cells are a mixture of small and large atypical cells. Fat necrosis and prominent karyorrhexis are frequently found.

Tumor cells express T-cell associated antigens (CD2, CD3, CD5, CD7) and frequently cytotoxic molecules (CD8, TIA-1, Granzyme B). TCR genes are clonally rearranged.[57]

Mycosis Fungoides/Sezary Syndrome

Mycosis fungoides (MF) and leukemic variant of the disease called Sezary syndrome (SS) represent different stages or manifestation of a primary cutaneous T-cell lymphoma. Patients present cutaneous plaques or nodules and/or generalized erythroderma. Tumor cells are predominantly small cells and medium-sized cells with cerebriform nuclei that infiltrate the epidermis and later appear in the circulation.

The immunophenotype of the tumor cells is consistent with mature, memory, helper T-cell. T-cell associated antigens (CD2, CD3, CD5) are co-expressed with the CD4 molecule in the majority of the cases. Only rare cases show cytotoxic or suppressor (CD4- CD8+) phenotype. The cutaneous lymphocyte-associated antigen (CLA) which is the ligand of vascular E-selectin (ALAM-1) is expressed in MF/SS. In cases, where MF/SS progress to large cell lymphoma the tumor cells may express CD30.

Peripheral T-Cell Lymphoma, Not Otherwise Specified

Peripheral T-cell lymphoma, not otherwise specified (PTCL-NOS) comprises a heterogeneous group of peripheral T-cell neoplasms not be subcategorized under distinct entities. Lymphomas like lymphoepitheloid lymphoma, pleomorphic T-cell lymphoma, T-zone lymphoma and immunoblastic T-cell lymphoma determined in Kiel classification are falls in this category. They exhibit a range of cytologic grade, nonetheless, all are clinically aggressive.

PTCLs are express T-cell associated antigens and most cases CD4 positive. Rare cases may express CD8 or cytotoxic molecules. Aberrant T-cell associated molecule expression like selective loss of one or more T-cell markers or absent of both CD4 and CD8 molecules may occur. The TCR genes are clonally rearranged in most cases.

Angioimmunoblastic T-Cell Lymphoma

Angioimmunoblastic T-cell lymphoma (AIL-T) is usually associated with generalized lymphadenopathy, fever, weight loss, skin rash and polyclonal hyperagammaglobulinaemia. Histologically, the nodal architecture of lymph node is diffusely effaced and characteristically a proliferation of small arborizing high endothelial venuls is present. Reactive follicles with germinal centers are usually absent. The lymphoid infiltrates in the lymph node composed of a mixture of small lymphocytes, immunoblasts, and atypical "clear" cells. Atypical cells may form small aggregates or sheets.

Tumor cells express T-cell associated antigens and CD4, however, polyclonal plasma cells, B-cells and histiocytes are characteristic constitutes of AIL-T as well. CD21 positive follicular dendritic reticulum cells surround post-capillary venules. Immunoblasts are frequently CD30 positive and EBV gene expression (LMP-1, EBERs) is commonly found in these cells. TCR genes are clonally rearranged in the majority of the cases.[59]

Anaplastic Large Cell Lymphoma T/0-Cell, Primary Systemic Type

Anaplastic large cell lymphoma (ALCL) is a group of lymphomas of T- or 0-cell phenotype with consistent CD30 expression. Tumor cells are large cells with pleomorphic often multiple nuclei and single or multiple prominent nucleoli. Tumor cells infiltrate predominantly sinusoidal and paracortical zones of lymph nodes.

Tumor cells, by definition, are CD30 positive and in the majority of the cases also express CD45 and epithelial membrane antigen (EMA). About 75% of the cases express T-cell associated antigens (CD3, CD43 and CD45RO) and CD4 molecule. In some cases, TIA-1, Granzyme B and Perphorin are expressed suggesting cytotoxic lymphocyte origin of these cases. The CD20 and other B-cell associated antigens are negative.

About half of the ALCLs carry the t(2;5)(p23;q35) nonrandom chromosomal translocation. In this translocation, the nucleophosmin (NPH) gene located on 5q35 and anaplastic lymphoma kinase (ALK) gene located on 2p23 is involved, which results in a chimeric product called p80. As a result of the translocation ALK gene is over-expressed. Monoclonal antibodies against ALK and p80 proteins are available that help in the immunhistochemical diagnosis of ALCLs. ALK positive ALCLs have a distinctly favorable clinical course comparing to ALK negative cases.[60]

Anaplastic Large Cell Lymphoma, T/0-Cell, Primary Cutaneous Type

In primary cutaneous anaplastic large cell lymphoma the skin is infiltrated with cohesive sheets of large cells with anaplastic morphology and CD30 positivity. Primary cutaneous anaplastic large cell lymphoma is different from systemic form of ALCL. In primary cutaneous anaplastic large cell lymphoma tumor cells express T-cell associated antigens and sometimes cytotoxic molecules but lack EMA and do not carry the t(2;5) translocation. The disease is more indolent than systemic ALCL and spontaneous regression may occur.[61]

References

1. Harris NL, Jaffe ES, Stein H et al. A revised European-American classification of lymphoid neoplasms: a proposal from the International Lymphoma Study Group. Blood 1994; 84:1361-92.
2. Harris NL, Jaffe ES, Diebold J et al. The World Health Organization classification of neoplastic diseases of the haematopoietic and lymphoid tissues: report of the Clinical Advisory Committee Meeting, Ariline House, Virginia, November 1997. Histopathol 2000; 36:69-86.
3. Jennings CD, Foon KA. Recent advances in flow cytometry: Application to the diagnosis of hematologic malignancy. Blood 1997; 90:2863-92.
4. Frizzera G, Wu CD, Inghirami G. The usefulness of immunophenotypic and genotypic studies in the diagnosis and classification of hematopoietic and lymphoid neoplasms. Am J Pathol 1999; 111(suppl.1):S13-39.
5. Chang KL, Arber DA, Weis LM. CD20: A review. Appl Immunohistochemistry 1996; 4:1-15.
6. Mason DY, Cordell JL, Brown MH et al. CD79a: a novel marker for B-cell neoplasms in routinely processed tissue smples. Blood 1995; 86:1453-9.

7. Inghirami G, Szabolcs MJ, Yee HT et al. Detection of immunoglobulin gene rearrengement of B cell non-Hodgkin's lymphomas and leukemias in fresh, unfixed and formalin-fixed, paraffin-embedded tissue by polymerase chain reaction. Lab Invest 1993; 68:746-57.

8. Krenacs L, Harris CA, Raffeled M et al. Immunohistochemical diagnosis of T-cell lymphomas in paraffin sections. J Cell Pathol 1996; 1:125-36.

9. Knowles DM. Immunophenotypic and antigen receptor gene rearrangement analysis in T cell neoplasia. Am J Pathol 1989; 134:761-85.

10. Spits H, Lanier LL, Phillips JH. Development of human T and natural killer cells. Blood 1995; 85:2654-70.

11. Boulland ML, Kanavaros P, Wechsler T et al. Cytotoxic protein expression in natural killer cell lymphomas and in alpha beta and gamma delta peripheral T-cell lymphomas. Am J Pathol 1997; 183:432-9.

12. Okuda T, Fischer R, Downing JR. Molecular diagnosis in pediatric acute lymphoblastic leukemia. Mol Diagnosis 1996; 1:139-42.

13. Pui CH, Raimondi SC, Head DR et al. Characterization of childhood acute leukemia with multiple myeloid and lymphoid markers at diagnosis and at relapse. Blood 1991; 78:1327-37.

14. Thompson AA, Talley JA, Do HN et al. Aberrations of the B-cell receptor B29 (CD79b) gene in chronic lymphocytic leukemia. Blood 1997; 90: 1387-94.

15. Dorfman DM, Pinkus GS. Distinction between small lymphocytic and mantle cell lymphoma by immunoreactivity for CD23. Mod Pathol 1994; 7: 326-31.

16. Zukenberg L, Medeiros L, Ferry J et al. Diffuse low-grade B-cell lymphomas. Four clinically distinct subtypes defined by a combination of morphologic and immunophenotypic features. Am J Clin Pathol 1993; 100: 373-85.

17. De Rossi G, Zarcone D, Mauro F et al. Adhesion molecule expression on B-cell chronic lymphocytic leukemia cells: malignant cell phenotypes define distinct disease subsets. Blood 1993; 81:2679-87.

18. Matutes E, Oscier D, Garcia-Marco J et al. Trisomy 12 defines a group od CLL with atypical morphology: Correlation between cytogenetic, clinical and laboratory features in 544 patients. Br J Haematol 1996; 92:382-8.

19. Matolcsy A, Inghirami G, Knowles M. Molecular genetic demonstration of the diverse evolution of Richter's syndrome (chronic lymphocytic leukemia and subsequent large cell lymphoma). Blood 1994, 83:1363-72.

20. O'Brien S, del Giglio A, Keating M. Advances in the biology and treatment of B-cell chronic lymphocytic leukemia. Blood 1995; 85: 307-18.

21. Huh YO, Pugh WC, Kantarjian HM et al. Detection of subgroups of chronic B-cell leukemias by FMC7 monoclonal antibody. Am J Clin Pathol 1994; 101: 283-9.

22. Lennert K, Tramm I, Wacker H-H. Histopathology and immuncytochemistry of lymph node biopsies in chronic lymphocytic leukemia and immunocytoma. Leuk Lymphoma 1991; 5(suppl):157-60.

23. Krenacs L, Himmelmann AW, Quintanilla-Martinez L et al. Transcripting factor B-cell-specific activator protein (BSAP) is differentially expresed in B cells and in subsets of B-cell lymphomas. Blood 1998; 92:1308-16.

24. Shivdasani RA, Hess JL, Skarin AT et al. Intermediate lymphocytic lymphoma: Clinical and pathologic features of a recently characterized subtype of non-Hodgkin's lymphoma. J Clin Oncol 1993; 11:802-11.

25. O'Briain DS, Kennedy MJ, Daly PA et al. Multiple lymphomatous polyposis of thegastrointestinal tract. A clinicopathologically distinctive foem of non-Hodgkin's lymphoma of B-cell centrocytic type. Am J Surg Pathol 1989; 13:691-9.

26. Banks PM, Chan J, Cleary ML et al. Mantle cell lymphoma. A proposal for unification of morphologic, immunologic, and molecular data. Am J Surg Pathol 1992;16:637-40.

27. Williams M, Swerdlow S, Rosenberg C et al. Characterization of chromosome 11 translocation breakpoints at the bcl-1 and PRAD1 loci in centrocytic lymphoma. Cancer Res 1992; 52:5541-56.

28. Whang-Peng J, Knutsen T, Jaffe ES et al. Sequential analysis of 43 patients with non-Hodgkin's lymphoma: clinical correlations with cytogenetic, histologic, immunophenotyping, and molecular studies. Blood 1995; 85:203-216.

29. Flenghi L, Ye BH, Fizzotti M et al. A specific monoclonal antibody (GP-B6) detects expression of the BCL-6 protein in germinal center B cells. Am J Pathol 1995; 147:405-11.
30. Korsmeyer SJ: Bcl-2 initiates a new category of oncogenes: Regulators of cell death. Blood 1992; 80:879-86.
31. Stroup R, Shieibani K: Antigenic phenotypes of hairy cell leukemia and monocytoid B-cell lymphoma: an immunohistochemical evaluation of 66 cases. Hum Pathol 1992; 23:172-7.
32. Dierlamm J, Pittaluga S, Wlodarska I et al. Marginal zone B-cell lymphomas of different sites share similar cytogenetic and morphologic features. Blood 1996; 87:299-307.
33. Zucca E, Bertoni F, Roggero E et al. The gastric marginal zone B-cell lymphoma of MALT type. Blood 2000; 96:410-9.
34. Finn T, Isaacson P, Wotherspoon A. Numerical abnormality of chromosomes 3, 7, 12 and 18 in low grade lymphomas of MALT-type and splenic marginal zone lymphomas detected by interphase cytogenetics on paraffin embedded tissue. J Pathol 1993; 170:335.
35. Auer IA, Gascoyne RD, Connors JM et al. t(11;18)(q21;q21) is the most common translocation in MALT lymphomas. Ann Oncol 1997; 8:979-85.
36. Willis TG, Jadayel DM, Du MQ et al. BCL10 is involved in t(1;14)(p22;q32) of MALT B cell lymphoma and mutated in multiple tumor types. Cell 1999; 96:35-45.
37. Pawade J, Wilkins BS, Wright DH. Low-grade B cell lymphomas of the splenic marginal zone: a clinicopathological and immunohistochemical study of 14 cases. Histopathol 1995; 27:129-37.
38. Matutes E, Morilla R, Owusu-Ankomah K et al. The immunophenotype of splenic lymphoma with villous lymphocytes and its relevance to the differential diagnosis with other B-cell disorders. Blood 1994; 83:1558-62.
39. Matolcsy A, Chadburn A, Knowles DM. De novo CD5-positive and Richter's syndrome-associated diffuse large B cell lymphomas are genotypically distinct. Am J Pathol 1995; 147:207-16.
40. Noorduyn LA, de Bruin PC, Van Heerde P et al. Relation of CD30 expression to survival and morphology in large cell B cell lymphomas. J Clin Pathol 1994; 47:33-7.
41. Rodriguez J, Pugh WC, Cabanillas F. T-cell-rich B-cell lymphoma. Blood 1993; 83:1586-9.
42. Moller P, Moldenhauer G, Momburg F et al. Mediastinal lymphoma of clear cell type is a tumor corresponding to terminal steps of B cell differentiation. Blood 1987; 69:1087-95.
43. Nador RG, Cesarman E, Chadburn A et al. Primary effusion lymphoma: a distinct clinicopathologic entity associated with the Kaposi's sarcoma-associated herpes virus. Blood 1996; 88:645-56.
44. Vasef MA, Medeiros LJ, Yospur LS et al. Cyclin D1 protein in multiple myeloma and plasmocytoma: an immunohistochemical study using fixed, paraffin-embedded tissue sections. Mod Pathol 1997; 10:927-32.
45. Pelicci P, Knowles DM, Magrath I et al. Chromosomal breakpoints and structural alterations of the c-myc locus differ in endemic and sporadic forms of Burkitt lymphoma. Proc Natl Acad Sci USA 1986, 83:2984-8.
46. Weiss L, Bindl J, Picozzi V et al. Lymphoblastic lymphoma: an immunophenotype study of 26 cases with comparision to T cell acute lymphoblastic leukemia. Blood 1986; 67:474-8.
47. Sheibani K, Winberg C, Burke J et al. Lymphoblastic lymphoma expressing natural killer cell-associated antigens: A clinicopathologic study of six cases. Leuk Res 1987; 11:371-7.
48. Kitchingman G, Robigatti U, Mauer A et al. Rearrangement of immunoglobulin heavy chain gens in T cell acute lymphoblastic leukemia. Blood 1985; 65:725-9.
49. Knowles DM: The human T-cell leukemias: Clinical, cytomorphologic, immunophenotypic, and genotypic characteristics. Hum Pathol 1986; 17:14-33.
50. Loughran T. Clonal disease of large granular lymphocytes. Blood 1993; 82:1-14.
51. Pelicci P, Allavena P, Subar M et al. T-cell receptor (α,β,γ) gene rearrangements and expression distinguish large granular lymphocyte/natural killer cells and T-cells. Blood 1987; 70:1500-8.
52. Imamura N, Kusunoki Y, Kawa-Ha K et al. Aggressive natural killer cell leukemia/lymphoma: report of four cases and review of the literature. Possible existence of a new clinical entity originating from the third lineage of lymphoid cells. Br J Haematol 1990; 75:49-59.
53. Kwong Y, Chan A, Liang R. Natural killer cell lymphoma/leukemia: pathology and treatment. Hematol Oncol 1997; 15:71-9.
54. Watanabe S. Peripheral T-cell lymphomas and leukemias. Hematol Oncol 1986; 4:45-58.

55. Nagatani T, Matsuzaki T, Iemoto G et al. Comparative study of cutaneous T-cell lymphoma and adult T-cell leukemia/lymphoma:clinical, histopathological and immunohistochemical analyses. Cancer 1990; 66:2380-6.
56. Emile JF, Boulland ML, Haioun C et al. CD5- CD56+T-cell receptor silent peripheral T-cell lymphomas are natural killer cell lymphomas. Blood 1996; 87:1466-73.
57. Murray A, Cuevas EC, Jones DB et al. Study of the immunohistochemistry and T cell clonality of enteropathy-associated T-cell lymphoma. Am J Pathol 1995; 146:509-19.
58. Salhany KE, Feldman M, Kahn MJ et al. Hepatosplenic gamma/delta T cell lymphoma: ultrastructural, immunophenotypic and functional evidence for cytotoxic T lymphocyte differenciation. Hum Pathol 1997; 26:674-85.
59. Kumar S, Krenacs L, Mederios J et al. Suncutaneous panniculitic T-cell lymphoma in a tumor of cytotoxic T lymphocytes. Hum Pathol 1998; 29:397-403.
60. Tobinai K, Minato K, Ohtsu T et al. Clinicopathologic, immunophenotypic, and immunogenotypic analyses of immunoblastic lymphadenopathy-like T cell lymphoma. Blood 1988; 72:1000-6.
61. Shiota M, Fujimoto J, Takenaga M et al. Diagnosis of t(2;5)(p23;q35)-associated Ki-1 lymphoma with immunohistochemistry. Blood 1994; 84:3648-52.
62. de Bruin P, Beljaards R, van Heerde P et al. Differences in clinical bechaviour and immunophenotype between primary cutaneous and primary nodal anaplastic large cell lymphoma of T-cell or null cell phenotype. Histopathol 1993; 23:127-35.

CHAPTER 4

Clinical Application of Immunophenotyping in Immunodeficiencies

Katalin Pálóczi

Deficiencies of immune responsiveness can occur as either primary or secondary disorders. Secondary immunodeficiencies may be acquired as a consequence of treatment with immunosuppressive agents, bone marrow or organ transplantation, nutritional deficiencies, and certain viral infections, the most notable of which is the human immunodeficiency virus (HIV).

Congenital (Primary) Immunodeficiencies

The primary immunodeficiency diseases are inherited defects of the immune system that are typically manifested at birth or soon thereafter, although some become evident later in life.[1] Frequent or unusually severe infections are the hallmark of immune system defects, the types of infections reflecting the nature of the defect. Autoimmune disorders and malignancies also occur more frequently in individuals with immunodeficiency diseases.[2]

To understand immunodeficiency diseases, it is important to comprehend the normal development and function of the cells belonging to the immune system. The immune system employs two broad lineages of cells react specifically with antigens: B cells and T cells. B lymphocytes are precursors of antibody secreting cells of the humoral immune system. T lineage consists of different subtype of cells including cells that mediate important immunoregulatory functions such as help or suppression, as well as cells involved in effector functions, such as the direct destruction of antigen bearing cells and the production of soluble products termed cytokines. The normal cellular events involved in B and T cell differentiation are reviewed in the Chapter 1. That review forms the basis of the discussion of the pathogenesis of the various immunodeficiencies.

Immunodeficiency diseases may involve either specific or nonspecific limbs of the immune system. Defects of phagocytic cells and deficiencies of the various complement components are examples of the latter category. Defects of specific immunity, both cellular and humoral, will be in the focus of this chapter, in which representative immunodeficiency diseases within the context of altered development and function of T and B cells are discussed (Table 1).

This approach began with the discovery that T and B cells represent separate lymphocyte lineages and that prototypic immunodeficiency syndromes could be created experimentally by depletion of the thymus-dependent and/or bursa-dependent lymphocytes. This hypothetical framework, outlined in Figure 1, provides a practical basis for consideration of the pathogenesis of immunodeficiency diseases. With the development of monoclonal antibodies (MoAbs) and flow cytometry, the identification of cells expressing certain markers (phenotypes) and the association of phenotypes with different state and function have become feasible. Although

Immunophenotypic Analysis, Second Edition, edited by Katalin Pálóczi. ©2005 Eurekah.com.

Table 1. Primary immunodeficiency diseases

Cellular and Combined Immunodeficiencies	Antibody Deficiency Diseases
Severe combined immunodeficiencies	***Congenital agammaglobulinemias***
Autosomal recessive SCID	X-linked agammaglobulinemia
X-linked SCID	X-linked agammaglobulinemia
SCID with ADA deficiency	with growth hormone deficiency
SCID with PNP deficiency	Autosomal recessive agammaglobulinemia
SCID with MHC Class II deficiency	***Common variable immunodeficiency***
Deficiencies in expression of CD3 molecules	***Selective antibody deficiencies***
Reticular dysgenesis	IgA deficiency
Immunodeficiency associated with	IgG subclass deficiencies
other defects	***Others***
T cell activation deficiencies	κ-chain deficiency
DiGeorge syndrome	Thymoma and hypogammaglobulinemia
Wiskott Aldrich syndrome	
Hyper IgM syndrome	

MoAbs proved to be useful in the clinical evaluation of immunodeficiency diseases, many of these diseases are phenotypically heterogeneous (Table 2). The immunophenotyping of cells alone may provide only limited information because often no correlation between T or B cell phenotype and function of these cells in vitro or in vivo exist.[3]

Cellular and Combined Immunodeficiency Diseases

In the last few years, significant advances have been made in the molecular definition of human primary immunodeficiencies. T cell immunodeficiencies have been attributed to mutations of the genes encoding the interleukin (IL)-2γ receptor, CD3ε and γ chains as well as the CD40 ligand, causing a form of severe combined immunodeficiency (SCID), functional T cell deficiency and defective immunglobulin (Ig) switching, respectively.[4,5]

Severe Combined Immunodeficiency

The autosomal recessive form of severe combined immunodeficiency diseases (SCID) includes those immunodeficiencies in which both cellular and humoral immunity are profoundly impaired. About 20% of patients with SCID present a phenotype characterized by the absence of T and B lymphocytes. Functional natural killer (NK) cells, however, can be detected. The classical form of SCID, formerly called 'Swiss-type agammaglobulinemia' results from genetic lesions that prevent normal development of both T and B cells. Another autosomal recessive SCID is the ataxia telangiectasia and related syndromes.

An X-linked form of SCID can also occur as a consequence of an isolated developmental failure of T cells, since B cells require T cell help for their normal growth and differentiation. Another basis for combined immunodeficiency is the inability to express class II genes of the major histocompatibility complex (MHC). Still other causes of functional aberrations affecting both T and B cells are found in isolated instances. Impaired cellular and humoral immunity can result from a variety of defects, a few of which have been rather precisely defined.

SCID with B Cells (X-Linked SCID)

X-linked SCID is characterized by defective T cell differentiation while B cell maturation is preserved. Patients usually do not have any mature T cells but have an increased number of

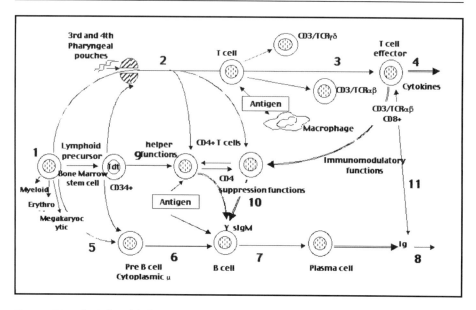

Figure 1. Hypothetical model of various immunodeficiency states. 1) Failure of T and B cell development (Reticular dysgenesia, SCID). 2) Failure of development of thymus (DiGeorge syndrome). 3) Deficiencies in expression of CD3 molecules. 4) T cell activation deficiencies. 5) Failure of maturation of stem cells into pre-B cells (thymoma and hypogammaglobulinemia). 6) Failure of maturation of pre-B cell into B cells (X-linked hypogammaglobulinemia). 7) Failure of maturation of B cells into plasma cells. 8) Hypercatabolism of Ig. 9) Reduced helper function of T cells. 10) Increase in suppressor activity of T cells. 11) Excessive loss of immunoglobulins and lymphocytes.

B cells. T cells are missing as a consequence of abortive intrathymic development, the epithelial thymus being virtually devoid of lymphocytes in affected male infants. B cells are present in normal numbers and they are capable of differentiating into antibody-producing plasma cells. As an autosomal recessive SCID, X-linked SCID can be treated by bone marrow transplantation. Recently, it was found that mutations in the gene encoding the IL-2 receptor (IL-2R) γ-chain account for X-linked SCID phenotype.[6]

IL-2 deficiency in humans is associated with abnormal T cell differentiation.[7] This suggests that the IL-2 receptor γ-chain may be part of another critical receptor for T cell differentiation. The IL-2R γ chain is constitutively expressed in hematolymphoid cells and is known to be a component of additional cytokine receptors including IL-4, IL-7, IL-9, IL-15 and also possibly IL-13 receptors. The important finding that mutations in the gene encoding the IL-2 receptor γ-chain profoundly affect T, NK and to a lesser extent B cell differentiation will be a source of further understanding of the mechanism of lymphoid differentiation.[4] The treatment of choice is bone marrow transplantation with an HLA-matched donor. With an HLA identical donor 95% survival with full immune reconstitution may be expected but it drops to 50-60% survival with varying degrees of hematopoietic reconstitution even in haploidentical transplants.

SCID with Adenosine Deaminase Deficiency

Adenosine deaminase (ADA) deficiency, inherited as an autosomal recessive trait, results in a lymphopenic form of SCID in homozygous individuals that is fatal in early childhood.[4] Phenotypes vary from very early onset disease with no residual ADA activity and profound

Table 2. Predominant abnormalities in selected primary immunodeficiency diseases

Name of Deficiency Syndrome	Specific Abnormality	Immune Defect	Susceptibility
Severe combined immune deficiency (SCID)	ADA deficiency	No T or B cells	General
	PNP deficiency	No T or B cells	General
	X-linked SCID, γc chain deficiency	No T cells	General
	Autosomal SCID DNA repair defect	No T or B cells	General
DiGeorge syndrome	Thymic aplasia	Variable numbers of T and B cells	General
MHC class I deficiency	TAP mutations	No CD8 T cells	Viruses
MHC class II deficiency	Lack of expression of MHC class II	No CD4 T cells	General
Wiskott-Aldrich syndrome	X-linked; defective WASP gene	Defective polysaccharide antibody reponse	Encapsulated extracellular bacteria
Common variable immunodeficiency	Unknown; MHC-linked	Defective antibody production	Extracellular bacteria
X-linked agammaglobulinemia	Loss Btk tyrosine kinase	No B cells	Extracellular bacteria, viruses
X-linked hyper IgM-syndrome	Defective CD40 ligand	No isotype switching	Extracellular bacteria
Selective IgA and/or IgG deficiency	Unknown; MHC-linked	No IgA synthesis	Respiratory infections
Phygocyte deficiencies	Many different	Loss of phagocyte function	Extracellular bacteria and fungi
X-linked lymphoproliferative syndrome	SAP mutant	Inability to control B cell growth	EBV-driven B cell tumors
Ataxia telangiectasia	Gene with PI-3 kinase homology	T cells reduced	Respiratory infections

The specific gene defect, the consequence for immune system, and the resulting disease susceptibilities are listed for some common and some rare human immunodeficiency syndromes. ADA= adenosine deaminase; PNP= purin nucleotide phosphatase; TAP= transporters associated with antigen processing; WASP= Wiskott-Aldrich syndrome protein; EBV= Epstein-Barr virus; NK= natural killer

lymphocytopenia to late onset disease associated with residual enzymatic activity and mild T cell lymphocytopenia. The deleterious effects of ADA deficiency on immune function are variable. Lymphocyte numbers are often very low but can be nearly normal in some ADA deficient individuals. Similarly, immunoglobulins of all isotypes may be grossly deficient, while other affected individuals may have nearly normal immunoglobulin levels.[1]

The immunodeficiency associated with ADA deficiency has been treated succesfully by bone marrow transplantation. However, when no matched donor is available there are now other options as bovine ADA coupled to polyethylene glycol (PEG-ADA) to increase stability and half-life and decrease immunogenicity given by three times weekly injection until somatic gene therapy are perfected.[3]

SCID with Purin Nucleoside Phosphorylase Deficiency

Purine nucleoside phosphorylase (PNP) deficiency is a rare disease also inhibiting proliferation of T cell precursors. The immunodeficiency is characterized by deficient T cell immunity and normal B cell function. It is often associated with neurological symptoms that remain poorly understood. PNP deficient individuals are often anergic and lymphopenic due to reduced numbers of circulating T cells. T cell responses to mitogens and antigens are reduced or absent. In contrast, the numbers of circulating B cells are normal or only slightly reduced. Serum immunoglobulins are also usually normal, as are antibody responses to blood group antigens, immunizations or natural infections. There is no specific replacement therapy for this enzyme deficiency. Bone marrow transplantation is at present the only feasible therapeutic approach.[1]

SCID with MHC Class I/II Deficiency

One rare form of combined immunodeficiency is called major histocompatibility compex (MHC) class I deficiency (bare lymphocyte syndrome). These patients often have normal number of T and B cells, but these fail to express classes I or II or both class antigens. The presumed pathogenesis is a defect of gene transcription within the cells.

Another autosomal recessive form of SCID is characterized by an inability to express class II molecules of the MHC. Normal numbers of T and B cells are generated in these individuals, but antigen-presenting cells either lack HLA-DR, HLA-DQ, and HLA-DP molecules entirely or express them in very low levels.[1] Thus macrophages, B cells, Langerhans' cells, and dentritic cells all fail to express MHC class II molecules. The defect extends to cells that do not constitutively express these molecules but that normally can be induced to do so.

The T cells present in affected individuals can be activated by mitogens or allogeneic lymphocytes, but they fail to express class II molecules following activation. The defective antigen specific T and B cell response in vitro and in vivo are caused by defective expression of MHC class II molecules.[9] This phenotype results from a defect in a transregulatory element of MHC class II gene expression.[4] Most patients with a deficiency in the expression of MHC class II molecules exhibit a partial but variable reduction in the number of CD4+ T cells, as expected from the known role of MHC class II molecules in CD4+ T cell differentiation.[4]

The variable clinical picture and the functional heterogeneity of T and B cells observed in this group of patients suggests that SCID with MHC class II deficiency might be heterogeneous disorders with different genetic basis.

Deficiencies in Expression of CD3 Molecules

Mutations affecting the genes encoding CD3γ and ε chains have recently been described in patients with immunodeficiencies. Interestingly, the absence of CD3γ protein did not abolish TCR-CD3 expression completely and 50% of T cells express TCR-CD3 complexes.[4] CD8+ T cell counts were reduced in these patients leading to the suggestion that the CD8 molecule interacts with CD3γ in the antigen recognition unit.[10] The occurrence of severe autoimmunity could point to abnormal intrathymic T cell selection although similar manifestations have been observed in other T cell immunodeficiencies.

The development of functionally immature T cells with variable phenotypic patterns has been observed in a few individuals with SCID.[11] Two infants were found to possess circulating lymphocytes with the immature thymocyte pattern of CD2 expression without detectable CD3, CD4, or CD8 expression.[12] Peripheral T cells with the intermediate thymocyte pattern of CD2, CD3, CD4, and CD8 expression have been observed in some infants with SCID, while still others were found to have an isolated deficiency of T cells with the CD4+ phenotype.[1,12]

Reticular Dysgenesia

Reticular dysgenesia is a rare congenital disorder that is characterized by impaired development of multiple cell lineages.[1] In addition to the early developmental failure of both T and B cell lineages, there is impaired growth and differentiation of myeloid cells in affected newborns. The disorder does not feature a complete developmental failure of hematopoietic stem cells; affected individuals produce erythroid and megakaryocytic cells in normal numbers. Successful treatment of an affected infant by bone marrow transplantation suggests that reticular dysgenesia may reflect inherent defects shared by myeloid and lymphoid cell precursors.[1]

Immunodeficiency Associated with Other Defects

T Cell Activation Deficiencies

In recent years, functional T cell immunodeficiencies have been described with abnormalities either in signal transduction or in cytokine production. So far, however, the molecular basis of these immunodeficiencies has not been elucidated. They include a case where T cell activation could not be induced through the T cell receptor (TcR) triggering but could be induced by direct activation of protein kinase C and/or calcium entry. A functional decoupling of the TcR-CD3 complex from phospholipase Cγ1 activation was postulated.[13] T cell deficiencies associated with a lack of production of IL-2 have been described in three patients with defective IL-2 gene transcription.[14-17]

DiGeorge Syndrome

DiGeorge syndrome is due to abnormal migration of the third and fourth branchial pouches with defective development of thymus. Detection of absence or extremely low number of T cells by immunophenotyping of peripheral blood mononuclear cells is helpful in the diagnosis of DiGeorge syndrome and related diseases. Hemizygosity is found in a clinically overlapping syndrome called velocardiofacial syndrome, which does not include abnormal development of the thymus. Further work is necessary to identify the genetic basis of the migration of these pouches.[4]

Wiskott-Aldrich Syndrome

Wiskott-Aldrich syndrome (WAS) is a complex X-linked immunodeficiency associated with thrombocytopenia, abnormal platelets, eczema, infections and progressive T and B cell immunodeficiency with a constant defect in the production of antibodies to polysaccharides. Patients tend to develop autoimmune manifestations and lymphomas.[4] The molecular basis of WAS remains unidentified. Recently, morphological and membrane abnormalities as well as signal transduction defects were described and reduced expression of CD43 by T lymphocytes was demonstrated.[18,19] The expression of CD23 on Epstein-Barr virus (EBV) transformed B cells has also been found to be reduced in patients with WAS.[20]

Hyper IgM Syndrome

Hyper IgM syndrome (HIgM) is characterized by defective production of Ig requiring a switch process, i.e., IgG, IgA and IgE, whereas IgM concentration is either normal or increased. Several lines of evidence indicate that the X-linked HIgM is not a B cell deficiency. Mayer et al has been postulated a defect in the T helper cells.[21] They showed that B cells from patients could be triggered to produce IgG and IgA in the presence of a Sezary cell line. Recently it was found that both the WAS disease locus and the gene encoding the CD40 ligand, a molecule expressed on activated T (and B cells), are mapped to the same genetic locus of chromosome X.

It is also suggested that the same molecular interaction between T cells and other CD40 expressing cells (such as B cells, epithelial cells and follicular dendritic cells) could be involved in the control of infections caused by opportunistic microorganisms.[4]

Predominantly Antibody Deficiency Diseases

In this section we consider those antibody deficiencies that feature abnormal B cell development and function in the presence of essentially normal T cell development. Although subtle functional deficits of T cell mediated immunity may sometimes occur, these are usually acquired ones. Bacterial infections are the most frequent complication of the antibody deficiencies. In some antibody deficiencies autoimmune disorders occur with greater frequency than in the general population. Apart from medical management of the infections and autoimmune disorders, antibody replacement therapy is currently the only effective treatment. Bone marrow transplantation has not yet been successfully employed to correct any of the primary antibody deficiencies in humans.[1]

Congenital Agammaglobulinemias

X-Linked Agammaglobulinemia

The X-linked agammaglobulinemia (XLA) is a convenient prototype for antibody deficiency syndromes. Affected boys have undetectable or very low serum levels of all the immunoglobulin subtypes. Boys with XLA are normal at birth. They usually start to have infections only after they have catabolized most of their maternally derived IgG antibodies. Although most of the infections in XLA patients are bacterial in origin, persistent viral and parasitic infections may also occur. The cellular basis of the antibody deficiency is arrested B cell development in the bone marrow (Fig. 2). While mature B cells, plasma cells and plasma cell precursors are deficient, pre-B cells containing intracytoplasmic μ chains are usually produced in normal numbers. Pre-B cells are present in the bone marrow while further development of the B series appears to be arrested and B lymphocytes in the blood are absent or very rare.

Most B cells display characteristically an immature phenotype as defined by cell surface markers. T cell development appears normal in XLA patients, and enumeration of T cell subpopulations usually reveals no significant abnormalities.[1,22,23]

Autosomal Recessive Agammaglobulinemia

The familial occurence of agammaglobulinemia in girls suggests that one form of congenital agammaglobulinemia may be inherited as an autosomal trait.[23,24] This is relative rare, and too few cases have been observed to document the exact mode of inheritance or to define the B cell defect.[1]

Common Variable Immunodeficiency

The common variable immunodeficiency (CVI) syndrome, which is not considered to be a single disease entity, is typically characterized by a paucity of antibody producing plasma cells, low levels of most or all immunoglobulin isotypes, and recurrent infections.[23] Most individuals with CVI have normal numbers of B cells, but they fail to undergo plasmacytic cell differentiation (Fig. 2). Helper CD4+ cells are also usually present in normal numbers. Germinal center formation by proliferating B cells may be extensive in CVI patients, as reflected by hypertrophy of the spleen, lymph nodes, and intestinal lymphoid tissues. This B cells overgrowth appears to reflect a normal proliferative response to antigen stimulation in the absence of normal feedback control by endogeneously produced IgG antibodies.[1] Possible pathogenetic

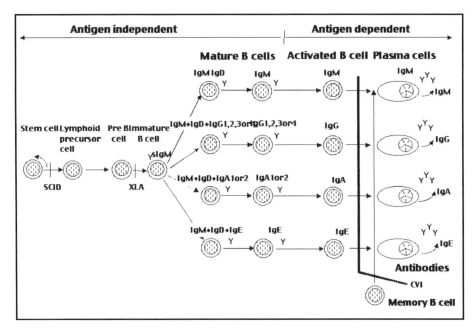

Figure 2. Scheme of B cell deficiencies. SCID= severe combined immunodeficiency; XLA= X-linked agammaglobulinemia; CVI= common variable immunodeficiency.

mechanisms include maturation defects restricted to B cells and suppression of immunoglobulin production by T cells. With increasing age many of these subjects develop defects of cell mediated immunity, detected as reduced numbers of T cells in blood and impaired proliferative responses to mitogens.[26]

IgA Deficiency

Isolated IgA deficiency is usually defined by very low level of serum IgA in the presence of normal or increased IgM and IgG. The clinical consequences of IgA deficiency range from severe systemic infections to a perfectly healthy state.[23] Most IgA-deficient patients never become aware of their antibody deficiency, having no more than the usual upper respiratory infections and occasional diarrhea. Others may have recurrent infections, interstitial malabsorption, allergic diseases, and autoimmune disorders such as rheumatoid arthritis, systemic lupus erythematosus, hemolytic anemia, and chronic active hepatitis. One reason for this remarkable spectrum of clinical manifestations may be the variability in the extent of the immunodeficiency. Another potential variable is the ability to replace IgA antibodies with IgM antibodies in the mucosal secretions in some IgA deficient individuals.[1]

The cellular defect in IgA deficiency is characterized by a failure of IgA B cells to undergo plasma cell differentiation (Fig. 3). The IgA genes appear to be normal, and the membrane-bound forms of IgA1 and IgA2 are expressed. The IgA bearing B cells produced by IgA deficient individuals appear to be arrested at an immature stage of development. Most of these IgA B cells still express IgM, as is normally the case in newborns but not in immunologically mature individuals.[1]

T cell development appears normal in most IgA deficient individuals, but T cells that can selectively suppress differentiation of IgA B cells have been found.[1]

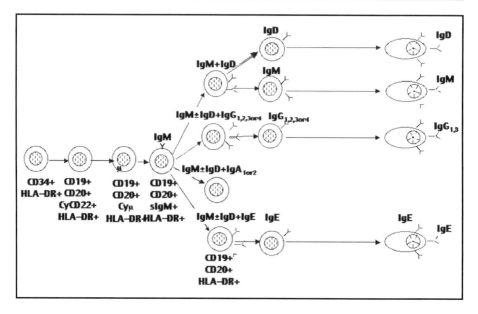

Figure 3. Cellular basis of IgA deficiency.

Selective Deficiencies of Other Immunoglobulin Isotypes

IgM deficiency is rare disorder that may be inherited as an autosomal recessive trait. Affected individuals have normal numbers of IgM and IgD bearing lymphocytes. The molecular basis for the selective inability to undergo IgM plasma cell differentiation is unknown, but laboratory data suggest abnormal reception of appropriate T cell help as one basis for selective defect in IgM B cell differentiation. IgG subclass deficiencies may involve one or more of IgG subclasses and can easily go undetected when total IgG is measured.

The diagnosis is usually made during the evaluation of individuals with recurrent bacterial infections who have no other obvious immunodeficiency. IgG subclass deficiencies have also been found in association with IgA deficiency. The immunological defect appears to be in the regulation of differentiation into IgG producing cells. Immunophenotyping is not helpful in diagnosis of IgG subclass deficiencies.[1]

Kappa Chain Deficiency

Immunoglobulin positive B cells with κ chains are prevalent in humans with a normal κ/λ ratio approximately 2:1. Rare individuals suffering from recurrent infections are selectively deficient in κ chain production. Cell surface κ chains are not detectable on peripheral B cells of these patients by immunophenotyping.[1,23]

Thymoma and Hypogammaglobulinemia

The association of thymoma and hypogammaglobulinemia in an immunodeficient individual was one of the first clues suggesting a role for the thymus in immunity.[1,26] Individuals with thymoma and hypogammaglobulinemia characteristically lack both pre-B cells in bone marrow and the cell of antigen dependent B cell proliferation in peripheral tissues. T cells are maintained in relatively normal numbers, even following removal of the whole thymus including the thymoma, although subtle deficits in cellular immunity are sometimes demonstrable.[1]

Phagocytic Cell Disorders

Chronic Granulomatous Disease

Chronic granulomatous disease (CGD) consists of a group of several disorders characterized by defective phagocyte function. Phagocytic cells in children with CGD ingest organisms normally but fail to kill them as a result of defects in the superoxide-producing pathway. This results in recurrent and chronic bacterial and fungal infections in affected individuals. All forms of the disease affect the various components of the nicotinamide adenine dinucleotide phosphate (NADPH) oxidase system of phagocytic cells. The NADPH oxidase is now to consist of four specific components, two of which are cytosolic: p47-phox and p67-phox, and two are membrane bound flavocytochrome b558 components: gp91-phox and p22-phox. Defects in any of these components give rise to CGD, with over 90% of all cases caused by defects in p47-phox and gp91-phox.[27,28]

Leukocyte Adhesion Deficiency

Leukocyte adhesion deficiency (LAD) is an autosomal recessive inability of leukocytes to adhere to cells such as the endothelium, inhibiting their ability to migrate to the sites of inflammation or infection. Defective expression of specific leukocyte surface integrins CD11/CD18 accounts for the clinical manifestations. Several families studied have specific mutations in their CD18 genes coding for the β2 integrin subunit. The CD11/CD18 molecules are adhesion antigens. CD11/CD18 are present on all leukocytes (monocytes, macrophages, lymphocytes, PMN and natural killer cells). Leukocyte functions mediated by CD11/CD18 are summarized in Table 3. Deficiency of the CD11/CD18 leukocyte adhesion molecules is a rare inherited disease presenting in the first 2 years of life and characterized by recurrent and often fatal bacterial infections.[27]

Reduced surface expression of CD11/CD18 is observed in specific granule deficiency is due to the absence of secondary and tertiary granules while in neonatal granulocytes there is a relative deficiency of these granules. Secondary and tertiary granules are the major intracellular storage sites for CD11b,c/CD18. Deficient CD11a/CD18 expression is also observed in certain lymphomas.[27] There are also acquired CD11/CD18 deficiency states such as ischemic-reperfusion injury and transplant rejection. In patients with burns, sepsis, hemodialysis, systemic lupus erythematosus and diabetes mellitus an increased expression of CD11/CD18 complex was observed.

Mannose-Binding Protein Deficiency

During the past few years it has become a prime suspect for unexplained immunodeficiency with a history of recurrent infection and inability of phagocytosis. This was subsequently found to be due to lack of a protein that bound to mannan, a mannose binding protein (MBP) which was absent or present in very low concentration in the blood of those with this impairment of phagocytosis. MBP is a major acute phase protein circulating in the blood in a wide range of concentration (0.1-50 mg/ml). Structural analysis shows that it resembles C1q by binding to ligands activating a serine protease and can then form a C3 convertase. Mutations in the MBP gene have been identified. These mutations apparently prevent MBP from polymerizing and performing its function. However, the existence of mutation in MBP, in 17% of Caucasians, is more frequent than the problems of immunodeficiency, and why this should be so has yet to be explained.[28]

Table 3.   Leukocyte functions mediated by CD11/CD18 antigens

Leukocyte Function	Relevant Heterodimer	Clinical Manifestation
Myeloid series		
Binding to iC3b	CD11b,c/CD18	Recurrent pyogen
Spreading/migration/chemotaxis	CD11a,b,s/CD18	infections
Aggregation	CD11a,b/CD18	Persistent neutrophilia
Adhesion to endothelium	CD11a,b,c/CD18	Poor granulocyte
Phagocytosis	CD11b/CD18	mobilization
Particle-induced oxidative burst and degranulation	CD11b/CD18	Poor wound healing
ADCC	CD11a,b,c/CD18	
Lymphoid series		Minimal
Antigen, mitogen, or alloantigen induced proliferation	CD11a/CD18	
NK, K and CTL	CD11a-c/CD18	
T and B aggregation	CD11a/CD18	
Adhesion to endothelium	CD11a/CD18	
Helper activity for in vitro Ig production	CD11a/CD18	

(References: 1,2,3,4,8,22)

Complement Defects

Inherited defects have been described for most components of the complement system. These are very rare giving rise to susceptibility to autoimmune disorders for the early component deficiencies (C2 and C4) and susceptibility to bacterial infections such as Neisseria meningitidis for the later components (C5-9). Of the alternative pathway components, X-linked properdin deficiencies have been described with defined mutations in three or four families worldwide. C1 esterase deficiency is the most common defect presenting angioedema and/or abdominal pain. Oedema can be severe leading to pharyngeal and laryngeal swelling, stridor, or even death. Therapy for acute attack of swelling involves supportive care and infusions of C1 inhibitor have been used. Chronic therapy usually involves treatment with androgenic steroids for increase protein synthesis from C1 inhibitor gene.[28]

Cellular Control Proteins

The most common deficiency of plasma membrane proteins to control complement activation is paroxysmal nocturnal hemoglobinuria, an acquired clonal defect of bone marrow stem cells results in failure of expression of all membrane proteins with a phosphoinositol glycan anchor. There is absence of two complement control proteins, DAF and CD59, which leads to abnormal complement activation, deposition on, and enhanced sensitivity to lysis of these erythrocytes. The disease is characterized by increased complement consumption primarily by the alternative pathway, on the surfaces of the abnormal bone marrow derived cells. Lysis of erythrocytes leads to the primary clinical manifestation, severe hemolytic anemia.[28,29]

Table 4. Lymphocyte phenotyping in primary immunodeficiency diseases

Disease	B Cells	T Cells
SCID	Variable positivity with CD19, CD20, CD21, CD9	a) CD3-,CD4-,CD8-,CD2- b) CD3-,CD4-,CD8-,CD38+ c) CD3+,CD4+,CD8+,CD38+
ADA deficiency	Absent	Absent
Bare lymphocyte syndrome	Normal numbers of CD19, CD20, CD21, sIgD/sIgM+ cells HLA-I/II Ags negative	CD3+, CD2+, CD4/CD45RA+, CD8/CD28+ HLA-I/II Ags negative
X linked agammaglobulinemia	CD19+, CD9+, CD20+-,CD21-, CyIgM+, sIgM-	CD3+, CD2+, CD4/CD45RA+, CD4/CD8 ratio usually normal
Common variable immunodeficiency	Normal CD19, CD20, CD21, sIgM	Normal CD3, CD2 Decreased CD4/CD8 ratio
Selective IgA deficiency	Normal CD19, CD20, CD21 sIgA/sIgM double + sIgA/sIgD double +	Normal CD3, CD2, CD4, CD8

(References: 1,2,3,8,9) Cy= intracytoplasmic; sIg= surface immunoglobulin

Use of Immunophenotyping in Clinical Evaluation of Patients with Primary Immunodeficiencies

Immunophenotyping of lymphocytes is an important component of the evaluation of immunodeficient patients (Table 4). In patients with hypogammaglobulinemia, B cells may be absent (in SCID, for example). Conversely, if B cells are present, they may not be functional, because of the number of circulating B cells does not necessary correspond to the quantity of immunoglobulin found in the serum.[3] Because CD4+ T cells are necessary for the function of B cells (as well as other limbs of the immune response), their presence is important. In SCID these cells (particularly mature T cells) are absent. This is manifest by an inability of the patient's cells to respond to foreign antigens by either immunoglobulin production or cell mediated immune processes that protect the host from overwhelming infections. Therefore, immunophenotyping of lymphocytes is important in the diagnosis of SCID. Immunophenotyping can be a potentially valuable tool in the prenatal diagnosis of SCID and in conjunction with enzyme determinations it has been successful in diagnosis SCID with adenosine deaminase deficiency.

In T cell deficiencies immunophenotyping may or may not be useful for diagnostic studies. The most important feature of these patients is the function of their lymphocytes. However, studies of subsets of CD4 or CD8 cells may help define the basis for demonstrated functional defect.[3] Immunophenotyping of various cellular populations in immunodeficiencies are valuable in diagnosis, treatment and understanding the pathomechanism of these diseases.

Human Immunodeficiency Virus Infection and Acquired Immunodeficiency Syndrome

The first cases of the acquired immune deficiency syndrome (AIDS) were reported in 1981 but it is now clear that cases of the disease had been occurring unrecognized for about 4 years before its identification. The disease is characterized by a susceptibility to infection with opportunistic pathogens or by the occurrence of an aggressive form of Kaposi's sarcoma or B-cell lymphoma, accompanied by a profound decrease in the number of CD4 T cells. As it seemed to be spread by contact with body fluids, and was early suspected to be caused by a new virus, and by 1983 the agent now known to be responsible for AIDS, called the human immunodeficiency virus (HIV), was isolated and identified. There are known to be at least two types of HIV —HIV-1 and HIV-2— closely related to each other. HIV-2 is endemic in West Africa and is now spreading in India. Most AIDS worldwide is, however, caused by the more virulent HIV-1.[29]

Pathogenesis of the Immune Deficiency

Infection with HIV produces a wide range of qualitative and quantitative immunological changes, including defects in T cell, NK cell, monocyte and B cell function. Prominent among these changes are those that affect the CD4+ T lymphocyte. Normally, the CD4+ T cell provides regulatory factors that enhance the function of many other cell types. It is believed that many of the defects in function seen in HIV disease in the other cells are linked to changes in CD4+ T lymphocyte number and function. There is a characteristic and progressive fall in the peripheral blood CD4+ T cell number that closely parallels the clinical deterioration in patients with HIV infection. How HIV produces this effect remains unclear although a number of mechanisms have been proposed. Which mechanism is most important in vivo remains unknown, although the loss of CD4+ T cell numbers and function is likely to be multifactorial (Table 5). CD4+ T cell dysfunction as well as depletion contributes to the immune deficit. CD4+ T cells from HIV-infected patients are defective in their ability to proliferate in response to soluble antigen and to induce immunoglobulin production from B cells.

Following primary infection there is a characteristic CD8+ lymphocytosis that persists until very late-stage disease: some of these CD8+ cells are HIV-specific cytotoxic T cells as well as nonspecific NK cells. The late stage decline in CD8+ cell numbers suggests that CD4+ lymphocytes have a part in maintenance of CD8+ cell numbers and function perhaps via secretion of IL-2 and other cytokines.

In contrast to infection of CD4+ T cells, HIV infection of cells of the monocyte/macrophage lineage does not result in significant cell death. Rather, persistent infection with the potential for sustained viral production is the rule and monocyte/macrophage lineage cells may act as an important reservoir of HIV.

Besides alteration in T cells polyclonal B cell activation is characteristic of HIV infection and is marked by hyperimmunoglobulinemia and persistent generalized lymphadenopathy. The immunoglobulin may show restricted mobility on routine electrophoresis, suggesting a limited idiotypic specificity. Much of the immunoglobulin is directed against HIV epitopes although neutralizing antibody to HIV is frequently not produced. Nodal and peripheral blood B cells from HIV-infected patients spontaneously produce immunoglobulin and express surface activation markers, implying they are in an activated state.[28]

Flow Cytometry for Detection of Lymphocytes in HIV+ Patients

In HIV infection, alterations in CD4 subsets have been examined in natural history studies and treatment trials to determine whether a particular cell population is preferentially lost in HIV infection or if response to therapy can be predicted by changes in CD4+ T cell subsets. It is clear that CD45RO, CD45RA or CD29 markers on CD4+ cells are not strong predictors

Table 5. Possible mechanisms of HIV-related immunodeficiency

CD4+ T lymphocytes
Directly cytopathic
Intracellular interaction between gp120 and CD4
Syncytium formation by cells expressing gp120
Binding to uninfected CD4+ cells
Induction apoptosis
Impaired IL-2 production
Impaired antigen and mitogen responses
Altered surface expression of IL-2R, T cell receptor and CD4
Macrophages
Reduced chemotaxis
Reduced monocyte dependent CD4 T cell proliferation
Rediced Fc receptor function
Reduced C3 mediated immune complex clearance
Reduced MHC II expression and impaired antigen presentation
Reduced intracellular killing
CD8+ lymphocytes
Reduced IL-2, IL-2R and IFN gamma production
Reduced cytotoxiv activity
Natural killer cells
Reduced cytokine production
Reduced cytotoxic activity
B lymphocytes
Polyclonal activation by HIV and other viruses
Increased IgG1, IgG3, IgA, IgE, reduced IgG2 and IgG4
Reduced neoantigen responses
Other
Infection of bone marrow precursors
Infection of thymocyte precursors

for disease progression or response to therapy. To date, CD4+ T cell levels alone, rather than any CD4+ subsets are the strongest indicator of HIV immunosuppression and progression to AIDS. Although CD4+ T cell levels are not the perfect surrogates for the effects of antiretroviral therapy they have been helpful for determining at least part of the therapy effect.[30]

CD8+ T cell subsets detecting by immunophenotyping have a multifaceted role in the immune defense system against HIV infection. In early HIV infection elevated CD8+ cells may actually predict slow disease progression as it will probably indicate that the immune system is effectively fighting the viremia. Only in the late stages of HIV infection and AIDS does this response wane.

Early observations of the changes in T cell subsets in HIV infection were reported as a percentage and absolute number of lymphocytes that are CD4+ or CD8+. Unfortunately, measuring the number of circulating CD4+ and CD8+ cells tells us nothing about the cells present in the tissues, where much of the HIV replication takes place. Because the percent of CD4+ cells can be measured with good precision and accuracy, it is believed that this is a better and more reliable measure of immunosuppression in HIV infection than absolute numbers. In fact, both percentage and absolute numbers of CD4+ T lymphocytes have similar predictive value in prognosis.

NK cells (CD16+) are reported to be decreased in HIV infection, particularly late in disease. NK activity is also decreased. However, the importance of enumeration of NK cells in the natural history of HIV infection is unclear. B cells have found to be decreased during

HIV disease despite observation that early after infection patients are often hypergamma-globulinemic.[28]

To determine whether the type of response in HIV infection is TH1, TH2 or TH0 some researchers have examined intracytoplasmic cytokines in the lymphocytes. This can now be done on a single cell basis using three-color (four-color) flow cytometry allowing identify both memory CD4+ and CD8+ T cells and one intracellular cytokine.

Detection of HIV-infected cells using flow cytometry is an exiting application. However, methods for measuring virus directly in plasma have found favor in clinical laboratories.

When available, viral load measurements should be evaluated in addition to CD4 cell counts. Normal or high levels of CD4+ cells generally indicate that the immune system is competent and has not suffered the consequences of viral infection. Elevated CD8+ cells may actually predict a good prognosis because it will most likely indicate that the immune system is trying to defend itself against the virus.

References

1. Cooper MD, Butler JL. Primary immunodeficiency diseases. In:Paul WE, ed. Fundamental Immunology. New York: Raven Press Ltd, 1989:1033-57.
2. Frenkel J, Neijens HJ, Den Hollander JC et al. Oligoclonal T cell proliferative disorder in combined immunodeficiency. Ped Res 1988; 24:622-27.
3. Nicholson KA. Use of flow cytometry in the evaluation and diagnosis of primary and secordary immunodeficiency diseases. Arch Pathol Lab 1989; 113:598-605.
4. Fischer A. Primary T-cell immunodeficiencies. Curr Opin Immunol 1993; 5:569-78.
5. Matsumoto S, Sakiyama Y, Ariga T et al. Progress in primary immunodeficiency. Immunol Today 1992; 13:4-5.
6. Noguchi MY, Rosenblatt HM, Filipovith AH et al. Interleukin-2 receptor γ chain mutation in X-linked severe combined immunodeficiency in human. Cell 1993; 73-147-57.
7. De Saint Basile G, Le Deist F, Caniglia M et al. Genetic study of a new X-linked recessive immunodeficiency syndrome. J Clin Invest 1992; 89:861-66.
8. Blaese RM. Development of gene therapy for immunodeficiency: Adenosine deaminase deficiency. Pediatr Res 1993; 33:549-55.
9. Fischer A. Severe combined immunodeficiencies (SCID). Clin Exp Immunol 2000; 122:143-149.
10. Arnaiz-Villena A, Timon M, Corell A et al. Brief report: Primary immunodeficiency caused by mutations in the gene encoding the CD3γ subunit of the T lymphocyte receptor. N Engl J Med 1992; 327:529-33.
11. Buckley RH, Gilbertsen RB, Schiff RI et al. Heterogeneity of lymphocyte subpopulations in severe combined immunodeficiency: Evidence against a stem cell defect. J Clin Invest 1976; 58:130-36.
12. Reinherz EL, Cooper MD, Schlossman SF. Abnormalities of T cell maturation and regulation in human beings with immunodeficiency disorders. J Clin Invest 1981; 68:699-705.
13. Chatila T, Wong R, Young M et al. An immunodeficiency characterized by defective signal transduction in T lymphocytes. N Engl J Med 1989; 320:696-702.
14. Chatila T, Castigli E, Pahwa R et al. Primary combined immunodeficiency resulting from defective transcription of multiple T -cell lymphokine genes. Proc Natl Acad Sci USA 1990; 87:10033-37.
15. Disanto JP, Keever CA, Small TN et al. Absence of interleukin 2 production in a severe combined immunodeficiency disease syndrome with T cells. J Exp Med 1990; 171:1697-704.
16. Weinberg K, Parkman R. Severe combined immunodeficiency due to a specific defect in the production of IL-2. N Engl J Med 1990; 322:1718-23.
17. Rijkers GT, Scharenberg JGM, van Dongen JJM et al. Abnormal signal transduction in a patient with severe combined immunodeficiency disease. Pediatr Res 1991; 29:306-9.
18. Molina IJ, Kenney DM, Rosen FS et al. T cell lines characterize events in the pathogenesis of the Wiskott-Aldrich syndrome. J Exp Med 1992; 176:867-74.
19. Remold-O'Donnell E, van Brocklyn J, Kenney DM. Effect of platelet calpain on normal T lymphocyte CD43: Hypothesis of events in the Wiskott-Aldrich syndrome. Blood 1992; 79:1754-62.

20. Simon HU, Higgins EA, Demetriov M et al. Defective expression of CD23 and autocrine growth-stimulation in EBV-transformed B cells from patients with Wiskott-Aldrich syndrome. Clin Exp Immunol 1993; 91:43-9.

21. Mayer L, Kwan SP, Thompson C et al. Evidence for a defect in 'Switch' T cells in patients with immunodeficiency and hyperimmunoglobulinemia M. N Engl J Med 1985; 314:409-13.

22. Fleisher TA, White RM, Broder S et al. X-linked hypogammaglobulinemia and isolated growth hormone deficiency. N Engl J Med 1980; 302:1429-34.

23. Waldmann TA. Immunodeficiency diseases: Primary and acquired. In: Samter M, ed. Immunological diseases. Boston/Toronto: Little, Brown and Company, 1988:411-64.

24. Hoffman T, Winchester R, Schulkind M et al. Hypogammaglobulinemia with normal T cell function in female siblings. Clin Immunol Immunopathol 1977; 7:364-71.

25. Rosen FS, Cooper MD, Wedgwood RJP. The primary immunodeficiencies. I. N Engl J Med 1984; 311:235-42.

26. Chapel H, Haeney M. Essentials of clinical immunology. Oxford: Blackwell Scientific Publications, 1984:37-65.

27. Arnaout A. Structure and function of the leukocyte adhesion molecules CD11/CD18. Blood 1990; 75:1037-50.

28. Forsyth K, Kinnon C, Levinsky RJ. Congenital immunodeficiencies. In: Bradly J, McCluskey J, eds. Clinical immunology. Oxford: Oxford University Press, 1997:3-27.

29. Janeway CA, Travers P, Walport M et al. Immunobiology. The immune system in health and disease. 4th edition. New York (USA): Churchill Livingstone, 1999:445-459.

30. Nicholson JKA, Mandy FF. Immunophenotyping in HIV infection. In: Stewart CC, Nicholson JKA, eds. Immunophenotyping. New York: Wiley-Liss, 2000:261-319.

CHAPTER 5

Immunophenotyping in Hematopoietic Stem Cell Transplantation

Katalin Pálóczi

B one marrow transplantation (BMT) is being increasingly used in clinical medicine and in many cases is the only treatment offering long-term cure for otherwise fatal diseases. In 1957 Thomas and coworkers from Seattle first showed that large quantities of bone marrow (BM) could be obtained and safely infused into humans.[1] Historically, this approach was pioneered, after early failures, using BM cells obtained from HLA-matched sibling donors as the source of the graft. These grafts are available to only one-quarter to one-third of suitable patients, and considerable effort has been exerted to find alternative donors for patients lacking a matched sibling donor. Approaches have included the use of hematopoietic stem cells obtained from matched unrelated donors (obtained through national and international marrow registries), partially mismatched related marrow donors, placental and umbilical cord blood and autologous marrow or peripheral blood (PB).[1]

In contrast to BM, where there is usually a single harvest of hematopoietic progenitor cells (HPC), the size of which is based on volume and/or nucleated cell numbers, multiple apheresis are routinely performed to collect peripheral blood progenitor cells (PBPC) due to the lower frequency of these cells in the peripheral circulation.[2] Measurements of committed progenitor cells by colony forming assays were unsuitable for this purpose, since they require 12-15 days to generate the results and therefore could not be used on a daily basis to determine whether collections should continue. This problem was largely resolved by identification of the CD34 antigen on HPCs, and the availability of monoclonal antibodies (MoAb) to this antigen, which could then be used for enumeration of these cells by flow cytometry.[3]

Use of CD34 for Identification of Hematopoietic Progenitors

The discovery, purification, and labeling of MoAb to an antigen known as CD34 made possible the enumeration of progenitor cells in the PB and BM using flow cytometry.[4,5] The CD34 antigen is found on a variety of early committed progenitor cells including, it is believed, the pluripotent hematopoietic stem cell (HSC) and stromal cell precursors. It is also present on endothelial cells in some tissues. Siena and coworkers (1991) were the first to report the use of flow cytometry for the measurement of CD34 HPC in order to determine the timing of optimum mobilization and whether the presence of CD34 on the progenitor cells correlated with engraftment potential.[6] In the clinical practice the percentage of CD34+ cells is calculated using a two-dimensional dot plot of CD34+ cells versus side scatter. An alternative approach is to use a multiparameter sequential gating strategy to define the specifically stained population.[7] In this case positive staining for CD34 and CD45, as well as characteristic light scatter properties are determined. The CD34+ population forms a cluster when displayed, on

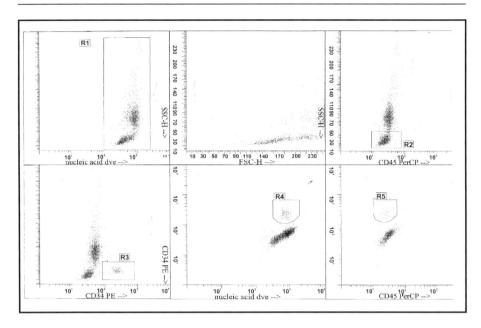

Figure 1. Peripheral blood stem cell apheresis product with three-color staining: nucleic acid dye/CD34/CD45.

a plot of CD45 staining versus side scatter that is distinct from that of lymphocytes, monocytes and neutrophils (Fig. 1).

There has been some controversy about the minimum dose of CD34+ cells required to achieve stable and sustained engraftment. The minimum number of CD34+ HPC is thought to be 2-5 x 10^6/kg recipient body weight.[6] That dose is, in most cases, easily obtained from BM and from small number of collections from mobilized PB of patients or normal donors.

Reconstitution of Hematopoiesis Posttransplant and Normal Hematopoiesis During Ontogeny

Normal adult hematopoiesis depends on the orderly development of hematopoietic cells within a specialized tissue, the bone marrow cavity. The blood-forming or hematopoietic system has been extensively studied for a variety of reasons. It represents the prototype self-renewing biological system in that large numbers of blood cells have to be produced daily in order to compensate for the loss of relatively short-lived mature blood cells. Investigations of the homing of transplanted hematopoietic cell into preconditioned recipients and the ability of transplanted cells to regenerate and maintain the hematopoietic system of an organism are very important in stem cell biology and hematopoietic stem cell transplantation (HSCT).

Hematopoiesis in Early Life

Hematopoiesis is first observed in 'blood islands' in the yolk sac.[8] Subsequently the liver, spleen and finally the bone marrow are colonised.[8-11] Based on avian experiments, the yolk sac does not contain precursors of adult hematopoiesis, which can first be found in the aorta-gonad-mesonephros (AGM) region at day 10 in gestation of mammalian development.[1,3,12-14] From the AGM region haematopoiesis moves to the fetal liver before reaching its final destination in the bone marrow (Fig. 2). In the adults, the bone marrow stem cells and circulating stem cell pools are in dynamic equilibrium, with over 98% of all committed progenitor cells in the marrow at any one time and very few stem cells circulating in the blood.[3]

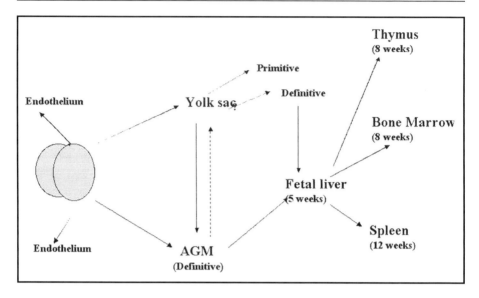

Figure 2. Models of ontogeny of mammalian hematopoiesis. Currently available evidences suggest the early appearance of hematopoietic cells both in the extraembryonic yolc sac and within the embrio proper, in the aorta-gonad-mesonephros (AGM) area. Definitive progenitor cells present in circulation colonize the liver at ~5-6 weeks and full development of liver hematopoiesis is established by 8-9 weeks and maintained about 12 weeks when bone marrow and spleen are colonized. Bone marrow is the exclusive site of hematopoiesis in the adult mice.

Phenotypic Characterization of Recovering Cells Posttransplant

The marrow microenvironment in the adult human serves as the site of normal hematopoiesis. Within this environment, hematopoietic progenitor cells lodge, proliferate and differentiate. In the setting of marrow transplantation, infused hematopoietic progenitor cells 'home' to the marrow and lodge. Furthermore, long term marrow cultures, early progenitor cells can be found preferentially associated with the stromal layer, too. These observations suggest that stem cells and early progenitors cells possess the ability to bind to marrow stromal components. The presence of 'homing receptors' on HSCs facilitates reconstitution of the BM from the blood following myeloablative therapy and haematopoietic progenitor cell reinfusion. Successful engraftment requires not only adequate numbers and quality of HSC but also appropriate localization of these cells within the bone marrow microenvironment (Fig. 3).[1,3,15,16]

HSCT is used to restore normal hematopoiesis following myeloablative and immunosuppressive chemotherapy or chemoradiotherapy. This process requires both lineage-committed progenitors, to effect early engraftment, and primitive progenitors, to effect long-term reconstitution (Fig. 4).[3,16,17] Although lineage committed progenitors are readily detected by in vitro cloning assays, there is no method that positively identifies the most primitive totipotent HSCs. Both primitive and lineage committed progenitor cells are known to be present within the CD34+ cell population of marrow which constitutes 1-5% of cells in adult BM.[3,17-19] Other recent publications confirmed the idea of CD34 negative earliest stem cells and present intriguing new data about progenitor cell development demonstrating the reversibility of CD34 expression on pluripotent hematopoietic cells.[18]

HSCT also requires the recipient to create a new immune system as stem cell divide and differentiate into lymphocyte effector cells (Fig. 5). Recovery of a competent immunological system capable of defending the marrow transplant recipient against pathogens is vital to the

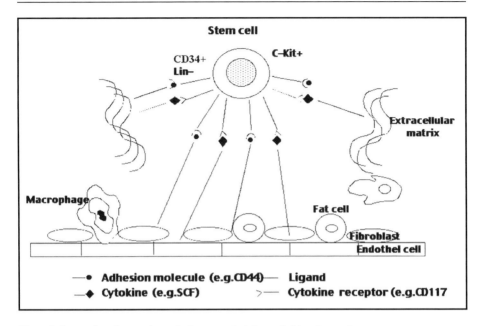

Figure 3. Stem cell and stromal matrix: hematopoiesis in a suitable microenvironment.

success of allogeneic HSCT. However, following allogeneic HSCT, recipients are characterized by a varying period of immunoincompetence.[19-21] The primary variables that contribute to posttransplant immunodeficiency include a lack of sustained transfer of donor antigen specific immunity, an impaired recapitulation of immunological ontogeny, and the effect of aging on thymic function. Additional factors that may contribute include donor-recipient histoincompatibility, the presence of acute and/or chronic graft versus host disease (GVHD), recipient age and anti-GVHD therapies.[21]

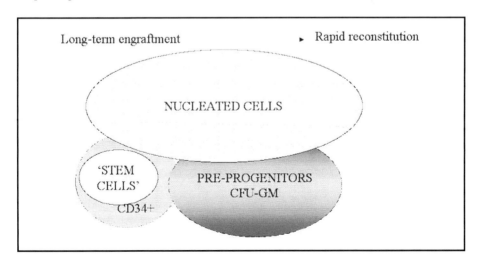

Figure 4. Hematopoietic cells in early and late engraftment. Among the transplantable nucleated cells there are stem cells showing CD34 negativity and preprogenitor cells showing CD34 antigen positivity responsible for late and early engraftment, respectively.

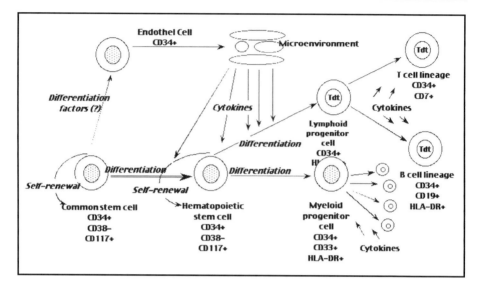

Figure 5. Hypothetical scheme of stem cell engraftment and differentiation.

Other additional factors including the nature of lymphocyte migration into lymphoid and nonlymphoid tissues following HSCT is currently a subject of scientific inquiry. There is increasing evidence that the capacity of lymphocytes to migrate into lymph nodes is compromised following HSCT, and that alterations in lymphocyte migration into lymphoid tissues contribute to delays in immune recovery.[22]

Cellular Reconstitution after Transplant

Leukocytes usually begin to reappear in the blood during the second or third week posttransplant. The newly derived neutrophils and monocytes seem to be able to carry out their most important functions as soon as they are generated. Such acquisition of full normal function, however, does not apply to T and B lymphocytes, some of whose functional activity is depressed long-term posttransplant.[20]

There are three main reasons for this delay in acquisition of lymphocyte function; first, it simply takes time for full functional development of lymphocytes to take place. These cells are undergoing ontogeny in a relatively adult allogeneic environment where, not only are there often minor or major histocompatibility differences between donor and recipient, but the amount of thymic epithelial tissue available for T lymphocyte education is markedly reduced compared with the situation obtaining during fetal lymphocyte development. A second major reason for the suppression of lymphocyte function is the administration of immune suppressive drugs to the recipient posttransplant with the primary aim of minimizing the severity of graft versus host disease (GVHD). The drug cyclosporin and especially methotrexate, that are used as prophylaxis for GVHD may result in the selective destruction of antigen-specific T lymphocytes that are stimulated in vivo by the presence of the specific antigen in the recipient. Third, GVHD selects the lymphoid system (besides the skin, liver and gut) as a target organ. Lymphoid hypocellularity and atrophy are characteristic histological hallmarks of moderate or severe GVHD.[23]

Kinetics and Function of Immune System Cells Posttransplant

Three factors govern the kinetics and function of immune system cells posttransplant: they are the rate of disappearance of host cells, the rate of appearance of cells derived from donor stem cells, and the persistence of mature (predominantly) lymphocytes infused in the donor inoculum.

Disappearance of Host (Recipient) Peripheral Blood Cells

With the initiation of the preparative regimen pretransplant, a rapid decline in peripheral blood white cells of host (recipient) origin begins. The rate of decline depends to some extent on the preparatory regimen utilized. Regimens employing busulfan[24] or dibromomannitol[25] are associated with a slower rate of decline than those utilizing total body irradiation.[25] The duration of the nadir of the white blood cell count is dependent on both the rate of disappearance of host cells and the rate of appearance of newly generated donor origin cells. Neutrophils present in the donor inoculum have little impact on circulating cell numbers in the recipient, although mature lymphocytes present in the donor graft have been shown to contribute functionally to recipient immunity posttransplant.[21,23]

T Cell Reconstitution after BMT into Adult Patients

Bone marrow derived HSCs in the normal process of differentiation home to the thymus, the major site of T cell differentiation. However, the thymus is not the only site of T cell development. T cell differentiation also occurs through extrathymic pathways in the gut mucosa, in the liver, and, at least in the case of murine T cell development, in the bone marrow as well.[26-28]

Over the last 10-15 years, immune reconstitution after marrow transplantation has been viewed as having some characteristic of the immune development in early life.[1,29] However, Storek et al provided phenotypic data suggesting marked discrepancies between T cell development in early life and after marrow grafting.[30] (a) The thymus in early life is well-developed; however, the thymus of an adult patient after transplant is naturally involuted and, furthermore, damaged by radiation, cytotoxic drugs or GVHD;[1] (b) Normal to markedly supranormal CD4+ T cell blood counts are typical for the late fetal, neonatal and infant period of life;[30] However, subnormal CD4+ T cell counts are generally detected for more than 1 year after marrow grafting into adult patients.[30,31] Markers of naivety (e.g., CD45RA) are infrequently expressed and antigens associated with memory (e.g., CD45RO) are frequently expressed on posttransplant T cells. Also in contrast to ontogeny, CD28 is not expressed on the majority of T cells, however, there is an increased percentage of T cells expressing CD11b, CD57, HLA-DR. Early activation markers like CD25 or CD69 are usually not expressed during the first year after transplant.[30,31] In contrast to CD4+ T cells, CD8+ T cell counts reach normal adult values within several months posttransplant resulting in CD4:CD8 ratios of <1. In contrast to their limited ability to produce naive CD4+ T cells, adult transplant recipients appear capable of producing naive CD8+ T cells. It is speculated that this could be due to extrathymic de novo CD8 T cell production.[26-28] Analysis of T cell receptor reconstitution by T cell receptor (TCR) Vβ rearrangements is a very useful method to determine the contribution of CD4 and CD8 subsets in recapitulation of T cell repertoire is also important for detection of host (endogenous) T cells and provides a new insight into thymic activity posttransplantation and may lead to strategies aimed at reactivating the thymus.[32]

The reconstitution of the recipient immune system requires the generation of new antigen specific T lymphocytes derived from the engrafted donor HSC. The generation of new T lymphocytes is dependent upon the function of the recipient thymus. In nontransplant settings, investigators have demonstrated an inverse correlation between age and the capacity of patients receiving chemotherapy to generate naive (CD45RA+CD4+) T lymphocytes.[21] Sustained cellular and humoral immunity in HSCT recipients requires the production of naive T lymphocytes capable of differentiating into antigen-specific T cells. Antigen specific T lymphocyte immunity is necessary for clinical control of DNA and RNA viral, protozoan, and fungal infections. Through their control of specific antibody production by B lymphocytes, T lymphocytes are also necessary for control of infections with encapsulated respiratory bacteria. Assessments of the transfer of donor T lymphocyte immunity have failed to demonstrate clinically significant transfer.[32,33]

Following HSCT, recipients are at high risk for infections with DNA viruses to which the recipients had pretransplant immunity. Both in vitro and in vivo function of T cells are impaired for 2 to 4 years posttransplant.[1,33]

B Cell Reconstitution after BMT into Adult Patients

Quantitative recovery of circulating B cells may be viewed as triphasic: (1) barely detectable B cell counts from the time of transplant until 3 to 6 months after transplant; (2) rapidly increasing B cell counts leading to supranormal level between 6-24 months after transplant; and (3) subsequent normalization probably over the following several years. Early B cells are larger and express CD21, sIgM and sIgD less frequently and CD71 (transferrin receptor) more frequently than normal adult B cells. Early posttransplant, all the stages of the B cell program (activation-proliferation-differentiation into Ig-producing cells) fail. This has been evidenced by B cell hyporesponsiveness to polyclonal stimulation.[1,31] In late survivals without chronic GVHD, activation, proliferation and IgM secretion gradually become normal. The only abnormality persisting for more than 1 year after transplant is insufficient production of IgG and IgA, secondary perhaps to a lack of isotype-switched (memory) B cells.[21,23,30,31] In patients with chronic GVHD, defects of activation, proliferation, and differentiation tend to persist beyond 1 year after transplant.[1,30]

Serum immunoglobin levels fall near or below the low normal limit during the first week after transplant. They then return to the normal adult range over the next several months to years: IgM levels normalize within weeks, IgG1 and IgG3 within months and IgG2, IgG4, and IgA levels normalize within years after grafting. Unfortunately, the quantity of serum immunoglobulins is an imperfect index of humoral immunity after BMT since the quality may be abnormal (the idiotype repertoire may be limited or not directed against microorganisms). Mono/oligoclonal antibodies and autoantibodies are frequently detected.[1,23,30,31] Antibody responses to polysaccharide antigens normalize only after several years even in patients without chronic GVHD.

Natural Killer Cell and Monocyte Number and Function

Large granular lymphocytes with natural killer (NK) activity constitute a major portion of lymphocytes repopulating the peripheral blood early after grafting. These cells are capable of producing cytokines and may play a role in regulating hematopoiesis, having been shown to both augment and suppress the growth of committed marrow progenitors. NK cells have also been suggested to play a role in the pathogenesis of acute GVHD, though this process is clearly initiated by donor derived T cells.[1,23,30,31]

Monocyte numbers in the blood return to normal rapidly posttransplant. Monocyte function after marrow transplantation has been less closely studied than T-and B-cell function, but all parameters thus far examined appear predominantly normal.[1,23,30,31]

Effects of GVHD on Immunological Reconstitution

Acute GVHD following histocompatible HSCT has little effect on the tempo of lymphoid reconstitution as measured by absolute lymphocyte count or the absolute numbers of CD3+ T lymphocytes. Transient depression of absolute lymphocyte count and CD3+ T cells

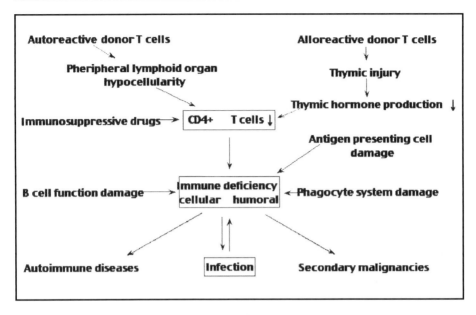

Figure 6. Components of the immune deficiency in chronic GVHD.

occurs following administration of anti-T lymphocyte immunosuppression, especially antithymocyte globulin (ATG) and anti-T lymphocyte (anti-CD3) monoclonal antibodies. The most apparent abnormalities due to GVHD are seen in chronic GVHD have a decreased capacity to develop antigen specific T lymphocyte response and to produce specific antibodies, particularly to polysaccharide antigens, while having an increased incidence of autoantibodies.[33] In vitro analysis of the cellular basis of the immunodeficiency present in patients with chronic GVHD has demonstrated a variety of cellular defects (Fig. 6). Decreases in the number and function of CD4+ T lymphocytes have been described, in addition to the presence of activated (HLA-DR expressing) CD4+ and CD8+ lymphocytes.[14,29] Defects in B lymphocyte function that result in a lack of normal responsiveness to T lymphocyte stimulation have been identified.[1,19,23,30,31] Thus, no single primary defect can explain the immunodeficiency associated with chronic GVHD.

Vaccination Posttransplant

Evidence suggests that immunity to specific antigens declines with time after transplant and may require periodic boosting, although probably not before two years or more from transplant.[29] Immunization on chronic GVHD patients with standard vaccines results in a poor immune response. Immunization with live virus vaccines is not recommended for fear of causing disease. Passive immune therapy with intravenous immunoglobulin has been shown to decrease the risk of Gram-negative sepsis and local bacterial infections, interstitial pneumonitis and should be utilized routinely in the transplant patients.

Mixed Chimerism

The allogeneic HSCT as it is currently defined includes three major components. These consist of conditioning regimens to both eradicate the underlying disease and suppress the host's immune system in preparation for the graft; infusion of the stem cell graft to rescue the recipient from otherwise lethal marrow toxicity of the conditioning regimen and to eliminate host resistance and residual leukemia via GVH reactions; and postgrafting immunosuppression or

T cell depletion to control GVHD and establish long-term graft-host tolerance. Complications related to the conditioning regimen are major limitation in the application of HSCT.

To reduce the risks of HSCT from toxic conditioning regimens, investigators have sought safer transplant programs. These regimens may result in stable mixed donor-host hematopoiesis instead of complete donor hematopoietic chimerism. Novel allotransplant studies embraces the concepts that the currently used intensive cytoreductive conditioning regimen can be replaced by nonmyelotoxic immunosuppression and that stem cell graft create their own marrow space through GVH reactions. Immunosuppression is divided into two parts, one aimed exclusively at host cells before transplantation, and the other directed at both host and donor cells after transplantation. The net effect is the establishment of mutual graft-versus-host tolerance as manifested by stable mixed donor-host chimerism.[33-35]

A great deal of effort has been expended in the twentieth century to understand how immune system distinguishes self from non-self. Fragments of mechanism have been unraveled, but there is still no clear understanding of the overall process. It seems likely that only a few of the millions of different clones of immunocytes recognize the foreign antigens in the transplanted cells. It is only these cells that need be eliminated. However, presumably because our understanding is incomplete, specific immune tolerance has not been achieved after allogeneic transplantation. If it can be achieved, cytoreduction may not be necessary for transplantation succeed.[1,31,34,35]

Multi-Tissue Potential of HSC

The hematopoietic system is not unique in its ability to generate large numbers of mature cells continuously throughout life. The intestinal epithelium, the skin epithelium and the male germline all share this property.[36] Stem cells have been purified from a number of organs and tissues including BM, liver, skin, gut, muscle and brain. The majority of stem cells reside in a quiescent or slowly cycling state and undergo cell division to self renew or differentiate into a transitional pool of rapidly dividing progenitor cells with limited self renewal. Progenitor cells undergo further lineage commitment and differentiation into the mature post-mitotic cells that constitute a specific tissue. At any times in this process, stem cells and their differentiated progeny may undergo apoptosis.

During the past year some groups reported that adult human stem cells have the ability to reconstitute human bone marrow even across the tissue border. They showed that stem cells from the adult brain retained the youthful ability to become several different kinds of tissues. Additionally, nonhematopoietic cells isolated from murine adult sceletal muscle seemed to exhibit even more BM reconstituting hematopoietic activity as compared to whole BM.[37] Recently, Keller (2001) has also proposed a pathway of pluripotent HSC maturation characterized by CD34 and c-Kit (CD117, stem cell factor receptor) expression suggesting that c-Kit[pos] pluripotent HSC might give rise to c-Kit[neg] cells and that these cells have multitissue potential (Fig. 7). Experiments are currently underway to determine the plasticity of the c-Kit[neg] cells. Progenitor and precursor cells in other organ systems, including germ, skin, muscle, liver, brain and gut, have been shown to express c-Kit, which suggests that the c-Kit maturation pathway may be the same for stem cells in other organ systems.[37]

Do these surprising observations warrant a rethinking of the current stem cell dogma? At this point, the answer is no—or at least, not yet.[36] In order to understand the rationale for this conclusion it is necessary to highlight again the role that clonal analysis has played in defining the hematopoietic stem cell. Simply stated, the experiments showing that a single cell population can yield hematopoietic as well as other tissues were not performed at the level of single cells. However, the biologic events that occur following HSCT seem to be determined by the presence of stem cells, progenitors, environmental factors, mature T cell and many soluble biological factors as well. A greater understanding of the interrelationship between the different cell types may result in improved transplantation results.

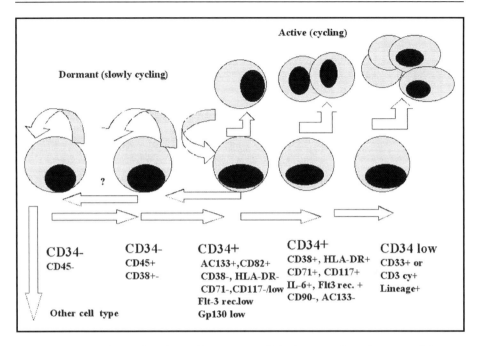

Figure 7. Summary of recent hypothesis of human stem cell development. The earliest stem cell (CD117 negative) is probably not tissue-determined, quiescent or slowly cycling with self-renewing capacity. Possibly through cell matrix interaction it can acquire a tissue specific phenotype and function. CD117 positive stem cells may have the potential to give rise to differentiated cells characteristic of other organs showing plasticity of stem cells. The expression of CD34 antigen is variable in the human bone marrow during maturation progress.

References

1. Thomas ED, Blume KG, Forman SJ. Hematopoietic cell transplantation. Malden: Blackwell, 1999.
2. Reiffers J, Goldman JM, Armitage JO. Blood stem cell transplantation. London: Martin Dunitz, 1998.
3. Stewart CC, Nicholson JKA. Immunophenotyping. New York: Wiley-Liss, 2000.
4. Civin CI, Loken MR. Cell surface antigens on human marrow cells: dissection of hematopoietic development using monoclonal antibodies and multiparameter flow cytometry. Int J Cell Cloning 1987; 5:267-288.
5. Tindle RW, Nichols RA, Chan L et al. A novel monoclonal antibody BI-3C5 recognizes myeloblasts and non-B non-T lymphoblasts in acute leukaemias and CGL blast crises, and reacts with immature cells in normal bone marrow. Leuk Res 1985; 9:1-9.
6. Siena S, Bregni M, Brando B et al. Flow cytometry to estimate circulating hematopoietic progenitors for autologous transplantation: comparative analysis of different CD34 monoclonal antibodies. Hematologica 1999; 76:330-333.
7. Sutherland DR, Keating A. The CD34 antigen: structure, biology and potential clinical applications. J Hematother 1995; 1:115-129.
8. Williams DA. Stem cell model of hematopoiesis. In: Hoffmann R, Benz EJ, Shattil SJ et al, eds. 2nd ed. Hematology. Basic principles and practice. New York: Churchill Livingstone, 1995:180-192.
9. Metcalf D. Ontogeny of the haematopoietic system: yolk sac origin of in vivo and in vitro colony forming cells in the developing mouse embryo. Brit J Haematol 1970; 18:279-296.
10. Lajtha LG. Stem cell concepts. Differentiation 1979; 14:23-34.
11. Broxmeyer HE. Self-renewal and migration of stem cells during embryonic and fetal hematopoiesis: important, but poorly understood events. Blood Cells 1991; 17:282-286.

12. Dieterlen-Lievre F, Godin I, Pardanaud L et al. Sites of hematopoietic stem cell production in early embriogenesis. In: Gluckman E, Coulombel L, eds. Ontogeny of hematopoiesis. Aplastic Anemia. Colloque INSERM 1995; 235:5-12.

13. Huang H, Auerbach R. Identification and characterization of hematopoietic stem cells from the yolk sac of the early mouse embryo. Proc Natl Acad Sci 1993; 90:10110-10114.

14. Medvinsky A, Dzierzak E. Definitive haematopoiesis is autonomously initiated by the AGM region. Cell 1996; 86: 897-906.

15. Cairns J. Mutation selection and the natural history of cancer. Nature 1975; 255:197-200.

16. Roy V, Miller JS, Verfaillie CM. Phenotypic and functional chracaterization of committed and primitive myeloid and lymphoid hematopoietic precursors in human fetal liver. Exp Hematol 1997; 25:387-394.

17. Raaphorst FM. Reconstitution of the B cell repertoire after bone marrow transplantation does not recapitulate human fetal development. Bone Marrow Transplant 1999; 24:1267-1272.

18. Nakamura Y, Ando K, Chargui J et al. Ex vivo generation of CD34(+) cells from CD34(-) hematopoietic cells. Blood 1999; 94:4053-4059.

19. Paloczi K. Clinical applications of immunophenotypic analysis. Austin: R.G. Landes Company, 1994.

20. Atkinson K. Reconstruction of the haemopoietic and immune system after marrow transplantation. Bone Marrow Transplant 1990; 5:209-226.

21. Parkman R, Weinberg KI. Immunological reconstitution following bone marrow transplantation. Immunol Rev 1997; 157:73-78.

22. Sackstein R. Disruption of lymphocyte homing to lymph nodes following bone marrow transplantation: Implications for immune reconstitution. Clin Immunol Newslett 1995; 15:144-147.

23. Witherspoon RP, Goehle S, Kretschmer M et al. Regulaton of immunoglobulin production after human marrow grafting: The role of helper and suppressor T cells in acute graft-versus-host disease. Transplantation 1986; 41:328-335.

24. Morgan M, Dodds A, Atkinson K et al. The toxicity of busulphan and cyclophosphamide as the preparative regimen for bone marrow transplantation. Br J Haematol 1991; 77:529-534.

25. Kelemen E, Masszi T, Reményi P et al. Reduction in the frequency of transplant-related complications in patients with chronic myeloid leukemia undergoing BMT conditioned with a new, nonmyeloablative drug combination. Bone Marrow Transplant 1998; 21:747-749.

26. Rocha B, Guy-Grand D, Vassalli P. Extrathymic T cell differentiation. Curr Opin Immunol 1995; 7:235-241.

27. Dejbakhsh-Jones S, Jerabek L, Weissman IL et al. Extrathymic maturation of $\alpha\beta$ T cells from hemopoietic stem cells. J Immunol 1995; 155:3338-3344.

28. Guillaume T, Rubinstein DN, Symann M. Immune reconstitution and immunotherapy after autologous hematopoietic stem cell transplantation. Blood 1998; 92:1471-1490.

29. Champlin R. Bone Marrow Transplantation. Boston: Kluwer, 1990.

30. Storek J, Witherspoon RP, Storb R. T cell reconstitution after bone marrow transplantation into adult patients does not resemble T cell development in early life. Bone Marrow Transplant 1995; 16:413-425.

31. Paloczi K. Immune reconstitution: An important component of a successful allogeneic transplantation. Immunology Letters 2000; 74:177-181.

32. McGreavely L, Fallen P, Travers P et al. Bone Marrow Transplant 2000; 25:S77.

33. Burt RK, Deeg HJ, Lothian ST et al, eds. Bone marrow transplantation Austin: R.G. Landes, Austin 1996:438-451.

34. Thomas ED. Semin Hematol 1999; 36(S7):95-103.

35. Weinberg K, Annett GM, Kashyap A et al. The effect of thymic function on immunocompetence following bone marrow transplantation. Biol Blood Marrow Transplant 1995; 1:18-23.

36. Papayannopoulou T, Lemischka I. Stem cell biology. In: Stamatoyannopoulos G, Majerus PW, Perlmutter RM et al, eds. The molecular basis of blood diseases. 3rd ed. Philadelphia: W.B. Saunders Co., 2001:1-24.

37. Brendel C, Neubauer A. Characteristics and analysis of normal and leukemic stem cells: current concepts and future directions. Leukemia 2000; 14:1711-1717.

38. Keller JR. Stem cell quiescence and activation. Mod Asp Immunobiol 2001; 1(5):217-220.

INDEX

A

Acute leukemia 29-32, 35, 38, 41-43, 54, 55

Acute lymphoid leukemia (ALL) 29, 31-38, 41-43, 49

Acute myeloid leukemia (AML) 29, 31, 32, 35-41, 43

Adenosine deaminase (ADA) 10, 63-65, 73

Adenosine deaminase (ADA) deficiency 63-65, 73

Adhesion molecule 1, 16, 18, 50, 53, 71

Adult T-cell leukemia/lymphoma (ATL/L) 56

AIDS 74, 75

Anaplastic large cell lymphoma (ALCL) 54, 58

Apoptosis 5-7, 10, 11, 52, 53, 75, 86

Ataxia telangiectasia 63, 65

Autosomal recessive agammaglobulinemia 63, 68

B

B cell 2-12, 15-19, 31, 33, 35, 36, 41, 43, 51, 52, 54, 62-70, 73-75, 84

B cell antigens 51

B cell deficiencies 67, 69

B cell reconstitution 84

Bare lymphocyte syndrome 66

Biphenotypic acute leukemia (BAL) 35, 41, 42

Bone marrow transplantation 64-68, 78

Burkitt's lymphoma (BL) 36, 54, 55

C

CALLA 4, 35

CD1 2, 9, 41, 49

CD2 2, 5, 9, 32, 35-37, 40-42, 49, 55-57, 66, 73

CD3 2, 9, 18, 19, 31, 32, 36, 41, 49, 55-58, 63, 64, 66, 67, 73, 84, 85

CD3 expression 66

CD4 2-4, 8-13, 19, 20, 32, 36, 38, 40, 41, 49, 55-58, 65, 66, 68, 73-76, 83, 85

CD5 2-4, 9, 32, 36, 41, 42, 49-57

CD7 2, 9, 31, 32, 36-38, 41-43, 49, 55-57

CD8 2, 3, 8, 10-13, 19, 20, 32, 36, 41, 42, 49, 55-57, 65, 66, 73-76, 83, 85

CD9 3, 4, 15, 36, 42, 73

CD10 2-4, 32, 33, 35-37, 41-43, 49-55

CD11 32, 71, 72

CD11/CD18 71, 72

CD11/CD18 deficiency 71

CD11b 2, 20, 31, 38, 40, 71, 72, 83

CD11c 2, 19, 20, 50-53

CD13 2, 14, 15, 20, 31, 32, 37-43, 49

CD14 2, 14, 15, 19, 32, 37, 38, 40, 42

CD15 2, 32, 35, 38-40, 42

CD16 10, 15, 18, 19, 49, 55-57, 75

CD18 71, 72

CD19 2-6, 18, 19, 31-33, 35-37, 39-43, 49-55, 73

CD20 2, 4, 31-33, 35, 36, 41, 42, 49-55, 58, 73

CD21 2-5, 41, 50-52, 55, 58, 73, 84

CD22 2, 5, 35, 36, 41, 49-55

CD23 2, 3, 5, 7, 15, 50-53, 67

CD24 5, 32, 42

CD25 2, 5, 8-11, 50, 53, 56, 83

CD29 74

CD30 10, 54, 57, 58

CD32 5, 15, 19